week

Molecular *and* Cellular Mechanisms *of* Toxicity

Molecular *and* Cellular Mechanisms *of* Toxicity

Edited by

Francesco De Matteis

Professor, Institute of Pharmacology,
University of Turin, Italy

Lewis L. Smith

Professor, MRC Toxicology Unit,
University of Leicester, England

CRC Press
Boca Raton   New York   London   Tokyo

Library of Congress Cataloging-in-Publication Data

Molecular and cellular mechanisms of toxicity / edited by Francesco de Matteis
  and Lewis L. Smith
      p.   cm.
    Includes bibliographical references and index.
    ISBN 0-8493-9229-2 (alk. paper)
    1. Molecular toxicology. 2. Genetic toxicology. I. De Matteis, Francesco.
  1932– . II. Smith, Lewis L.
  RA1220.3.M65   1995
  615.9--dc20   95-16810
                                                                             CIP

PREFACE

Studies of mechanisms of toxicity are of paramount importance in toxicology, as they influence our thinking across a wide spectrum of toxicological interests from the more basic to the applied. Apart from the obvious significance of these studies in clarifying the pathogenesis of the various toxic syndromes, elucidation of mechanisms of toxic responses helps the biochemist and physiologist to obtain a better grasp of the normal processes and provides the new concepts on which risk assessment can be scientifically based.

Mechanistic studies in toxicology were pioneered by John Barnes and his colleagues at the Medical Research Council Toxicology Unit at Carshalton and have remained a prominent feature of the work carried out by the Unit. A meeting was organized jointly by the Medical Research Council and the British Toxicology Society to coincide with the move of the Unit to its new site in the recently built Centre for Mechanisms of Human Toxicity at the University of Leicester. This book is based on the contributions made to that meeting and contains a number of short reviews by scientists, most of whom have either been members of the Unit or have been closely associated with it at some time. It is intended to highlight points of growing interest in the field of mechanistic toxicology, as well as to provide a more permanent tribute to the achievements of the MRC Toxicology Unit.

We have divided the content into three sections. The first considers selected aspects of molecular mechanisms, including selectivity of toxic agents and repair processes in the nervous system, toxicity of oxygen, fibers, and aflatoxins. The second part discusses, more specifically, the interactions of carcinogens with DNA and other targets and their relevance to both molecular dosimetry of exposure and development of cancer. The final part is concerned with cellular and genetic aspects and includes treatment of some of the most recent and rapidly developing problems in toxicology, such as apoptosis and the importance of genetic polymorphism, as well as studies of gene expression by *in situ* hybridization.

Francesco De Matteis and *Lewis L. Smith*
Turin and Leicester

CONTRIBUTORS

Professor W. Norman Aldridge
Robens Institute of Health & Safety
University of Surrey
Guildford, England

Dr. R.C. Brown
MRC Toxicology Unit
University of Leicester
Leicester, England

Dr. Gerald M. Cohen
MRC Toxicology Unit
University of Leicester
Leicester, England

Dr. J.M.G. Davis
Institute of Occupational Medicine
Edinburgh, Scotland

Professor Francesco De Matteis
Institute of Pharmacology and
  Experimental Therapy
University of Turin
Turin, Italy

Dr. David Dinsdale
MRC Toxicology Unit
University of Leicester
Leicester, England

Dr. Peter B. Farmer
MRC Toxicology Unit
University of Leicester
Leicester, England

Dr. Howard O. Fearnhead
MRC Toxicology Unit
University of Leicester
Leicester, England

Dr. Michael F.W. Festing
MRC Toxicology Unit
University of Leicester
Leicester, England

Professor Marcello Lotti
Istituto di Medicina del Lavoro
Universitá degli Studi di Padova
Padua, Italy

Dr. Ian Lauder
Department of Pathology
University of Leicester
Leicester, England

Dr. Marion MacFarlane
MRC Toxicology Unit
University of Leicester
Leicester, England

Professor P.N. Magee
Jefferson Cancer Institute
Thomas Jefferson University
Philadelphia, PA, USA

Dr. Margaret M. Manson
MRC Toxicology Unit
University of Leicester
Leicester, England

Professor Donald J. Reed
Department of Biochemistry and
  Biophysics
Oregon State University
Corvallis, OR, USA

Professor Rolf Schulte-Hermann
Institut für Tumorbiologie-
  Krebsforschung
University of Vienna
Vienna, Austria

Professor Lewis L. Smith
MRC Toxicology Unit
University of Leicester
Leicester, England

Dr. Xiao-Ming Sun
MRC Toxicology Unit
University of Leicester
Leicester, England

CONTENTS

MECHANISMS OF TOXICITY

Chapter 1
Aspects of Selectivity and Mechanisms in Neurotoxicity 3
W.N. Aldridge

Chapter 2
Injury and Repair in Neurotoxicology 19
Marcello Lotti

Chapter 3
Toxicity of Oxygen 35
Donald J. Reed

Chapter 4
Aflatoxins 69
Margaret M. Manson

Chapter 5
Experimental Studies with Fibers 85
J.M.G. Davis and R.C. Brown

BIOMONITORING AND MOLECULAR INTERACTIONS
IN CARCINOGENESIS

Chapter 6
Biomonitoring and Molecular Dosimetry of Genotoxic Carcinogens 101
Peter B. Farmer

Chapter 7
The Relationship of Molecular Interactions
with DNA to Cancer: Scientific Developments 125
P.N. Magee

GENETIC AND CELLULAR ASPECTS IN TOXICOLOGY

Chapter 8
The Study of Genes and Gene Expression by *In Situ* Hybridization:
A Potentially Useful Approach in Toxicology 147
Ian Lauder

Chapter 9
Genetic Variation in Human and Laboratory Animal Populations
and Its Implications for Toxicological Research and Human
Risk Assessment 165
Michael F.W. Festing

Chapter 10
Mechanisms of Cell Death, with Particular Reference to Apoptosis 185
**Gerald M. Cohen, Marion MacFarlane, Howard O. Fearnhead,
Xiao-Ming Sun and David Dinsdale**

Chapter 11
Apoptosis: Some Aspects of Regulation and Role in Carcinogenesis 207
Rolf Schulte-Hermann

INDEX 219

Part I
Mechanisms of Toxicity

Chapter 1

ASPECTS OF SELECTIVITY AND MECHANISMS IN NEUROTOXICITY

W. N. Aldridge

CONTENTS

I. Introduction 3

II. Delivery of the Intoxicant 5

III. Triorganotins 5
    A. Intramyelinic Vacuoles 6
    B. Neuronal Necrosis 6
    C. Structure-Activity 6
    D. Conclusions 8

IV. Pyrethroids 8
    A. Structure-Activity 9
    B. Dose-Response 9
    C. Conclusions 10

V. Organophosphorus-Induced Delayed Neuropathy 11
    A. Neuropathy Target Esterase Hypothesis 11
    B. Recent Observations 12
    C. Reflections 12

VI. Conclusions 13

References 14

I. INTRODUCTION

No subdivision of toxicology into specialties such as neurotoxicology existed when the Medical Research Council Toxicology Unit was established in early 1947. After the 1939-45 war, an intense synthetic program was undertaken in industry to exploit organophosphorus compounds as pesticides (Saunders, 1957; O'Brien, 1960; Heath, 1961; WHO, 1971). It was not unex-

0-8493-9229-2/95/$0.00+$.50
© 1995 by CRC Press Inc.

pected that a study of the action of organophosphorus compounds was one of the first research areas undertaken by the Unit.

Although it was appreciated that small molecular weight chemicals could be powerful tools in the study of organized biological systems (Bernard, 1875), this attitude was adopted only by a few in academic physiology and the emerging discipline of biochemistry. Peters (1963; 1969), investigating the mechanisms of action of arsenicals and fluoroacetate and the subsequent rational development of therapeutic agents, pointed the way forward. Thus research began in the Unit on the organophosphorus compounds using pharmacological and histological approaches and on the kinetics of their interaction first with the cholinesterases and then with esterases in general. Thus began a scientific policy, still operating, of research in mechanistic toxicology to provide the background essential for rational decision taking about practical problems arising from exposure of humans and animals.

In neurotoxicology the central problem in those early days (and also today) was to provide explanations for the selective vulnerability of cells in the nervous system to exposure to chemicals. As the specific toxicity of organophosphorus compounds caused by inhibition of acetylcholinesterase was understood, research on other mechanisms of toxicity gradually led to the general acceptance that specific clinical signs or biological effects result from interaction(s) of a chemical with specific molecular component(s) of particular cells.

From 1947 onward many chemicals which cause changes in the structure or function of the nervous system have been studied in the Unit (Table 1.1). Of these I have selected three — triorganotin compounds, pyrethroids, and organophosphorus compounds — to illustrate the development of concepts and mechanisms in neurotoxicology. The whole process from exposure to the appearance of clinical signs of toxicity consists of five distinct stages (Aldridge, 1981):

1.   Exposure and penetration into the organism
2.   Delivery of the intoxicant to its site of action
3.   Interaction of the intoxicant with the target cell and macromolecular components of it
4.   Early consequences of the primary interaction (3) defined in histopathological, physiological, biochemical, etc. terms
5.   Clinical symptoms, signs, and syndromes of toxicity

The first three stages are chemical-related events, and the last two are biology-mediated.

The nervous system is unique in many different features, such as the complex relationship between different cell types both inside and outside the nervous system, its energy requirements, the inability of adult neurones to divide, and the presence of long cellular processes. Nevertheless, many of the systems of stages 1 and 2 concerned with the delivery of the intoxicant to its site of action are the same for toxicity in the nervous system and in other organs. The blood-brain

TABLE 1.1
Some Neurotoxic Chemicals: MRC Toxicology Unit, 1947–1993

Chemical	Selected references
Acrylamide	Fullerton and Barnes, 1966; Edwards, 1975
Carbon disulphide	Caroldi et al., 1987
1,3-Dinitrobenzene	Romero et al., 1991; Ray et al., 1992
Methylmercury	Magos, 1982; Magos and Butler, 1976; Magos et al., 1985
Organophosphorus compounds	See text
Pyrethroids	See text
2,4,5-Tribromimidazole	Verschoyle et al., 1984
Trialkyl-tins and -leads	See text

barrier has special properties, but the general lipophilic nature of many neurotoxins ensures their ready uptake from the bloodstream.

II. DELIVERY OF THE INTOXICANT

The three classes of chemicals — triorganotins, pyrethroids, and organophosphorus compounds — are highly lipophilic and slightly soluble in water and thus pass readily through cell membranes. The latter two classes are uncharged; the triorganotins can exist in acid conditions as positively charged ions, but in physiological media they circulate as the lipophilic chloride and hydroxide.

Enzymic systems involving oxidation, ester hydrolysis, and group transfer lead to detoxification. Bioactivation systems in the liver, kidney, and other organs convert biologically inactive chemicals to active intoxicants, for example by converting $P = S$ to $P = O$ (Gage, 1953; Aldridge and Barnes, 1952), oxidation of the methyl group of one o-methyl group in tri o-cresyl phosphate followed by rearrangement to form a saligenin phosphate (Eto et al., 1962), and the oxidative release of an ethyl group from tetraethyltin (and tetraethyllead) to form the corresponding triethyl species (Cremer, 1958, 1959). Although the nervous system was thought to be inactive in oxidative bioactivation systems, recent research with more sensitive analytical techniques has indicated that they are present (Friedberg et al., 1990; Minn et al., 1991; Chambers, 1992). The total concentration of cytochrome P450s in the brain is low, but if they are located in one or a few cell types, then their concentration may be rather high.

III. TRIORGANOTINS

Research in the toxicity of the triorganotins became relevant when they were under development as industrial biocides and particularly as fungicides (van der Kerk, 1976). After an approach by van der Kerk, the Unit undertook to take a preliminary look at their toxicity.

A. INTRAMYELINIC VACUOLES

After administration of triethyltin salts, rats exhibit generalized weakness, dragging of the hind limbs, depression of the respiratory rate, and decrease in body temperature (Magee et al., 1957). Histological examination of the brain showed many spaces in white matter; these were subsequently shown to be water containing sodium and chloride, characteristic of extracellular fluid. Electron microscopy established that the water accumulated between the myelin sheaths (intramyelinic vacuoles) and nowhere else. This was therefore a highly specific lesion and was present soon after administration of triethyltin (Jacobs et al., 1977). A significant increase in water in rat brain derived from wet and dry weight measurements cannot be demonstrated for 9 to 12 hours. However, if after the administration of triethyltin the fall in temperature is prevented by keeping the rats in a raised environmental temperature, then an increase can be seen almost immediately (Lijinsky and Aldridge, 1975). Signs of poisoning originate from an increased cranial pressure; if the rats recover, the water content of the brain decreases, and there is no residual pathology. Triethyltin also produces this highly specific and reversible lesion in both the central and peripheral nervous systems in many other species, including humans.

B. NEURONAL NECROSIS

It has been reported (Barnes and Stoner, 1958) that the signs of poisoning by trimethyltin are different from those of triethyltin. Instead of becoming quiet, as after triethyltin, rats show tremors, hyperexcitability, aggressive behavior, weight loss, and convulsions. These observations were unexplained for 20 years until it was shown that in such rats bilateral and symmetrical neuronal necrosis occurs involving the hippocampus (largely sparing the Sommer sector), pyriform cortex, amygdaloid nucleus, and the neocortex (Brown et al., 1979). These signs of poisoning and histopathology have been seen in several other species of experimental animals and in humans. Therefore, in contrast to the reversible lesion observed after triethyltin, trimethyltin produces an irreversible neuronal loss in specific areas of the central nervous system. If both compounds are administered together, both lesions can be detected (Seawright et al., 1984).

C. STRUCTURE-ACTIVITY

The unexpected difference in the biological response of the mammalian nervous system to triethyltin and trimethyltin has been extended (Table 1.2). If compounds are synthesized with successive substitution in trimethyltin of methyl by ethyl, both types of lesion are produced. Thus, triorganotin compounds can be divided into several classes causing: (1) neuronal necrosis only, (2) neuronal necrosis and intramyelinic vacuoles, (3) intramyelinic vacuoles only, and (4) neither lesion. These results reinforce the view that independent mechanisms are involved in the production of these two lesions.

Since the poisoning episodes with phenolic hexachlorphane (Kimbrough and Gaines, 1971), it has been shown that other acidic substances also cause

TABLE 1.2
Structure-Activity Relationships for Compounds Causing Neuronal Necrosis or Intramyelinic Vacuoles

Tin and Lead Compounds

	Neuronal necrosis	Intramyelinic vacuoles	Mitochondria (B/A)	Ref.
4-Coordinate				
(1) Trimethyltin	+	–	290	1, 2
Trimethyllead	+	–	850	1, 2
Triethyllead	+	–	640	1, 2
(2) Dimethylethyltin	+	+	32	2, 3
Methyldiethyltin	+	+	33	2, 3
Methylethyl-n-propyltin	+	+	11	2, 3
Trimethyltin/triethyltin	+	+	–	12
(3) Triethyltin	–	+	15	1, 2
Tri n-propyltin	–	+	3.6	1, 4
Diethylphenyltin	–	+	1.6	5, 6
Ethyldiphenyltin	–	+	0.4	5, 6
5-Coordinate				
(4) (2-[(Dimethylamino)methyl] phenyl)diethyltin	ND	–	0.01	5, 6
(2-[(Dimethylamino)methyl] phenyl)methylphenyltin	ND	–	0.05	5, 6

Acidic Phenols and Phenylimino Compounds

	Neuronal necrosis	Intramyelinic vacuoles	Mitochondrial uncoupling	Ref.
(5) 2,4-Dinitrophenol	ND	–	+	7
Pentachlorophenol	ND	–	+	7
(6) 2,2'-Methylenebis[3,4,6-trichlorophenol] (hexachlorophane)	ND	+	+	7, 8
3,5-Diiodochloro salicylamide	ND	+	+	9
2'-Chloro-2,4-dinitro-5',6-di(trifluoromethyl) diphenylamine	ND	+	+	10
2,4-Dinitro-N-(2,4,6-tribromophenyl)-6-(trifluoromethyl) benzenamine	ND	+	+	11

Note: B/A is the ratio of the concentrations of the compound causing 50% of the maximum effect on ATP synthesis (B) and the Cl⁻/OH⁻ exchange (A); ND = not examined. References: (1) Aldridge et al., 1977; (2) Aldridge and Brown, 1988; (3) Aldridge and Street, unpublished; (4) Barnes and Stoner, 1958; (5) Verschoyle and Aldridge, unpublished; (6) Aldridge et al., 1981; (7) Lock, 1976; (8) Kimbrough and Gaines, 1971; (9) Lock, 1977; (10) Lock et al., 1981; (11) Lier and Cherry, 1988; (12) Seawright et al., 1984.

intramyelinic vacuoles (Table 1.2). The acidic group may be phenolic or in others a phenyl imino CNHC group. Not all acidic phenols are active.

All of the chemicals discussed in the above two paragraphs affect the energy conservation system in mitochondria. The acidic compounds cause classical uncoupling defined as an ability to reduce the phosphorylation of ADP while maintaining the consumption of oxygen. This activity is achieved by the carriage of protons across membranes and the consequent discharge of the H^+/OH^- gradient essential for the synthesis of ATP. The trialkyl-tin and -lead compounds induce a Cl^-/OH^- exchange across mitochondrial membranes, and also they inhibit the ATP synthase reaction by binding to one of the components of the system. Relative activity in bringing about these two effects varies from compound to compound (Selwyn, 1976; Aldridge et al., 1977; Aldridge and Brown, 1988).

D. CONCLUSIONS

Much research on the biological properties of triorgano-tin and -lead compounds and uncoupling agents indicates that they are rather specific and do not interact with many proteins. Their lipophilic properties and solubility in water also ensure that they are widely distributed, and current available information suggests that the cellular specificity in the central nervous system to cause the two lesions is not due to different cellular distribution (but see Toggas et al., 1992). Any generally applicable hypothesis about mechanisms must be speculative.

The production of intramyelinic vacuoles could result from the disruption of an energy-requiring physiological system to keep myelin sheaths together, but wet with extracellular fluid. The discontinuities in the structure-activity relationships make it unlikely that any one property can explain their action. Perhaps a cell has to be affected by two actions at the same concentration of agent, i.e., the ability to transport chloride across membranes together with a reduction of energy requirements of the cell. Properties specific to the cell affected cannot be excluded, e.g., the presence of chloride in oligodendroglia and Schwann cells. If this were a reasonable hypothesis for the trialkyltins, then their chloride-carrying property would have to be matched by the phenols and imino compound acting as carriers of sodium across membranes.

Neuronal necrosis is caused by triakyltins and trialkylleads with a relatively high activity in facilitating the Cl^-/OH^- exchange and could be associated with effects on systems in which chloride movement is linked to other transports, e.g., γ-aminobutyrate. Such primary effects need not necessarily be in the cells showing the lesions; influence from other cells removing inhibition could reduce the energy stores of the damaged cells and compromise their viability.

IV. PYRETHROIDS

During the period when Elliott was synthesizing new pyrethroid structures at Rothamstead to increase biological activity and stability to light, the Unit agreed to have a preliminary look at their toxicity. When decamethrin arrived (now called deltamethrin) and rats were dosed, it was seen that the signs of poisoning were different from those observed after other pyrethroids (Barnes and Verschoyle, 1974).

A. STRUCTURE-ACTIVITY

Examination of a variety of pyrethroid structures showed that they could be classified into two groups, those producing the same syndrome as deltamethrin and those resembling permethrin and cismethrin (Verschoyle and Aldridge, 1980). The two syndromes in rats were: (1) T (tremor)-syndrome, aggressive sparring, sensitivity to external stimuli, fine tremor progressing to gross whole-body tremor, and prostration with raised temperature; (2) CS (choreoathetosis-salivation)-syndrome, pawing and burrowing, salivation, coarse whole-body tremor, sinuous writhing, and clonic seizures. In general these syndromes correlated with the absence or the presence of an α-cyano group in the molecule, respectively. Rarely, a mixed syndrome was seen, i.e., a T-syndrome with some salivation.

Pyrethroids act on sodium channels and delay the closing of the sodium channel activation gate thus causing a prolongation of the sodium tail current (Narahashi, 1974; Vijverberg and van den Bercken, 1982). This prolongation was expressed as a time constant and, in the presence of pyrethroids, ranged from 13 to 1770 msec depending on their chemical structure (Vijverberg and van den Bercken, 1990); T-syndrome pyrethroids increased the tail current by >100 msec and those causing the CS-syndrome by <300 msec. This correlation is most interesting, but leaves unanswered the reasons for the rather abrupt change of syndrome with pyrethroid structure and a time constant of >300 msec.

Recent research has raised the possibility that this discontinuity in the structure-activity relationships is due to another action of pyrethroids. Deltamethrin also inhibits a neuronal voltage-dependent chloride channel at low concentrations (Forshaw et al., 1993; Forshaw and Ray, 1990). Further exploration of this finding and its structure-activity relationships may tell us if the discontinuity discussed above is due to two actions of CS-syndrome pyrethroids with an α-cyano group, in contrast to one action for those causing the T-syndrome (Forshaw et al., 1993).

B. DOSE-RESPONSE

The doses of pyrethroids required to cause poisoning by the oral route are much larger than those given by intravenous injection. No doubt this is due to factors involved in their delivery to the nervous system (detoxification, non-specific binding, etc.). Dose-response relationships at the target are fundamental and, when known, reduce the difficulties of prediction from animals to humans. For pyrethroids there is, at present, no method of measuring their concentration at the sodium channel, but measurements of average concentration in the nervous system is an approximation which excludes the effects of delivery systems. Intravenous injection with its rapid delivery to the nervous system gives a better indicator of "intrinsic toxicity" than other routes of administration.

Using intraperitoneal injection as an experimental procedure to ensure a time-dependent increasing concentration of pyrethroid in the nervous system, different signs of poisoning by deltamethrin appear at increasing times after injection. These signs are associated with different concentrations of parent pyrethroid in the brain; as the concentration falls, the recovery phase begins and the signs of poisoning disappear (Rickard and Brodie, 1985). Under these conditions the signs of salivation, tremor, and choreoathetosis occur at concentrations in brain of parent deltamethrin of approximately 0.1, 0.2, and 0.4 nmol/ g brain, respectively. If a lower dose only elicits some of these symptoms, then the tissue concentration achieved is the same as that shown above for that particular symptom. A similar study has been made relating dose, symptoms, and increases in cyclic GMP; both classes of pyrethroid causing the T- and CS-syndromes are associated with large increases in cyclic GMP in the cerebellum. Since other substances also cause this, the response is probably secondary to the intense muscular activity (Brodie and Aldridge, 1982; Lock and Berry, 1981).

The glucose content of areas of rat brain changes after administration of deltamethrin. Significant increases in glucose content occur in most with the exception of the cerebellum (Cremer et al., 1980). Glucose utilization is the converse, with highly significant increases in the cerebellum and less or no increase in other areas (Cremer et al., 1980). The generalized (except in the cerebellum) increase in brain glucose is probably a consequence of the 60% increase in blood glucose, and increase in glucose utilization in the cerebellum is due to the intense activity there as shown by the increase in cyclic GMP. Few pyrethroids have been studied in such detail as deltamethrin.

The rapid and quantitatively large responses of metabolism of different areas of the brain may now be studied by positron emission tomography (PET) and magnetic resonance imaging techniques. It is now known that the effects of certain substances (even cell death) can be greatly influenced by external stimuli which switch on the metabolic activity of different areas of the brain (Romero et al., 1991; Ray et al., 1992). There has been a long interest in the MRC Toxicology Unit in energy metabolism in the brain (Cremer et al., 1981; Cunningham and Cremer, 1981). During this research, initially stimulated by the finding that trialkyltin compounds were powerful inhibitors of mitochondrial energy conservation systems, one of the first observations was made (Cremer, 1964) of the importance of glutamate in brain intermediary metabolism.

C. CONCLUSIONS

Much more research is necessary before we can reach an understanding of the structure-activity relationships in the neurotoxicity of this class of compounds. A wide range of pyrethroid structures is now available, and DDT and other diaryl compounds (Holan et al., 1971) must be accommodated. A new pyrethroid, etofenprox, does not contain an ester group (Yoshimoto et al., 1989) and unlike other pyrethroids has a low toxicity for fish. With such broad structural requirements, presumably for interaction with sodium channels, it is interesting that the many chiral compounds show such large changes in activity and toxicity.

More understanding of the mode of action of pyrethroids is required so that unambiguous proof is available that effects such as paresthesia in humans and lesions in the peripheral nervous system are the result of their biological actions via changes in the operation of sodium channels in nerve conduction. The weight of evidence at present is that this is the case (Aldridge, 1990; Vijverberg and van den Bercken, 1990).

V. ORGANOPHOSPHORUS-INDUCED DELAYED NEUROPATHY

The acute toxicity caused by inhibition of acetylcholinesterase and the mechanism of interaction of organophosphorus compounds with esterases has been extensively reviewed (Koelle, 1963; Ballantyne and Marrs, 1992; Aldridge and Reiner, 1972).

The first occurrence of chronic delayed neuropathy was the poisoning episode observed in the U.S. in 1930 which was due to the consumption, during Prohibition, of illicit liquor contaminated with tri o-cresylphosphate. During research undertaken to identify the etiological agent, the hen was shown to be an excellent experimental model for the disease (Smith et al., 1930). Little progress occurred in understanding the mechanism of this disease, and it remained a curiosity until three cases were reported in staff working in the pilot plant of a pesticide manufacturer (Bidstrup et al., 1953). The chemical being synthesized was Mipafox® (N,N′-diisopropyl phosphorodiamidofluoridate), which was a direct inhibitor of esterases; following this episode, Mipafox and other direct inhibitors were shown to cause delayed neuropathy in the hen (Barnes and Denz, 1953); equally important, not all organophosphorus inhibitors were neuropathic. These observations opened a new window of opportunity for research on the mechanism of this delayed neuropathy.

A. NEUROPATHY TARGET ESTERASE HYPOTHESIS

From a conviction that organophosphorus compounds were specific inhibitors of esterases *in vivo*, the working hypothesis was put forward that inhibition of an esterase was essential for the development of the disease and that inactive compounds would not inhibit this esterase (Aldridge et al., 1969). Using both *in vivo* and *in vitro* tests, an esterase with such characteristics was detected and named neuropathy target esterase (NTE). Screening of many compounds showed that those compounds causing delayed neuropathy in the hen had inhibited

NTE by 70 to 80% and those compounds not causing the disease had not inhibited NTE by this amount (Johnson, 1969, 1974); this led to the "NTE(70-80%)-hypothesis." Further testing of this hypothesis revealed that some compounds, e.g., phosphinates, carbamates, and sulfonyl fluorides, while inhibiting NTE, did not cause delayed neuropathy. The latter compounds differed from those previously used to develop the first hypothesis in that the inhibited esterase could not age. Aging is the loss of one of the groups attached to the phosphorus in the inhibited esterase; phosphates, phosphonates, and phosphoramido compounds could, theoretically, lose an attached group, but the previous three groups of compound could not. This led to the "NTE(70-80%-aging)-hypothesis," which has been found to apply for a large range of structures (Clothier and Johnson, 1979, 1980; Johnson 1975a and b, 1982). The most powerful evidence that NTE was the relevant esterase in the initiation of delayed neuropathy was the observation that those compounds which inhibited NTE and *did not age* protected hens against the development of the disease by the subsequent administration of normally toxic compounds *which could age* (Johnson and Lauwerys, 1969; Johnson, 1970). Protection lasted over the period required for the inhibited NTE either to reactivate (after carbamates) or for new enzyme to be resynthesized so that 70 to 80% inhibition could be obtained by the subsequent dosing of a neuropathic compound (Aldridge, 1986).

B. RECENT OBSERVATIONS

Those compounds such as phenylmethanesulfonyl fluoride which are protectors when given before a neuropathic compound have now been shown to be promoters when given after it (Lotti et al., 1991; Pope and Padilla, 1990). Thus, hens dosed with a dose of a compound sufficient to inhibit NTE *in vivo* by only 30 to 40% will develop the disease when subsequently given a promoting agent.

Phosphoramido compounds which cause delayed neuropathy have recently been shown to do so without aging as judged by the fluoride-reactivation test (Johnson and Safi, 1992). The fluoride-reactivation test consists of incubating the esterase preparation with a high concentration of potassium fluoride; those inhibited esterases with both groups attached to the phosphorus atom are reactivated, whereas those which have lost one group and are negatively charged are not.

For many years it has been known that it is difficult to produce delayed neuropathy in rats and in young chickens. It is now known that the same dose which does not cause the disease in these animals does so if promoters are administered later (Moretto et al., 1991, 1992).

C. REFLECTIONS

Whereas the "NTE(70-80%-aging)-hypothesis" has much evidence in favor of it, the new evidence is difficult to assimilate. Besides initiation by inhibition of NTE, other actions affect the neuropathic activity of organophosphorus compounds *in vivo*; possibly two actions are necessary.

No answer to these inconsistencies is possible at the moment, but there are several unexplained facts:

1. The observation that the phosphoramidates cause delayed neuropathy without aging depends on the reliability of the fluoride-reactivation test. Unambiguous chemical proof of aging has only been provided for a few compounds.
2. The aging of inhibited NTE occurs with great facility with a wide variety of chemical structures. For some of these, the released group is attached to a site presumably adjacent to the catalytic center. This latter reaction is probably unique to NTE.
3. The physiological function of NTE is unknown.
4. What is happening during the silent period after inhibition of NTE and the appearance of signs of delayed neuropathy (14 to 30 days) is unknown. Signs of the disease appear when much of the NTE activity has returned.

VI. CONCLUSIONS

The usual approach in the study of mechanisms of toxicity is to try to establish the nature of a single target, the modification of which is necessary for the initiation of toxicity. This single-target approach has proved inadequate in chemical carcinogenesis, and it is now accepted that more than one event must take place. The three classes of chemicals chosen for discussion each produces different lesions in the nervous system of several mammalian species; this presumably means that the system affected is generally present. For all three examples, organophosphorus-delayed neuropathy, CS-syndrome by pyrethroids, and trialkyltin-induced intramyelinic vacuoles, no mechanism involving a single target seems likely. Maybe two targets are required — either an attack on two targets in one cell or an attack on different targets in different cells. Tentative suggestions have been made.

Practical problems can focus attention on the need for changes in direction of research. Although the growth of knowledge indicates the necessity for the existence of the specialty of neurotoxicology, from a toxicological point of view there is merit in such research being conducted in close collaboration, if not in the same institution, where other aspects of toxicity are being studied and a wide variety of disciplines are represented.

Progress in mechanistic research involves initial experiments based on some preconceived idea (or prejudice) followed by hypothesis, anomalous results, new techniques, new or modified hypothesis, and so on. Benefiting from the ability of organic chemists to synthesize new structures, examination of dose-response and structure-activity relationships can be informative; if the dose at or near the target cell is known, such relationships can be a powerful tool in the generation of new hypotheses or the dropping of inadequate ones. Almost all major difficulties in solving mechanistic problems are due to ignorance of the mechanisms of physiological control.

It is a pleasure to acknowledge the research carried out on neurotoxicological problems by staff and visitors in the MRC Toxicology Unit, which has provided the hard information for this short appreciation. I regret that space limitations have meant that research on many subjects by many scientists has not been discussed.

REFERENCES

Aldridge, W. N., Mechanisms of toxicity: new concepts are required in toxicology, *Trends. Pharmacol. Sci.*, 2, 228, 1981.

Aldridge, W. N., The biological basis and measurement of thresholds, *Ann. Rev. Pharmacol. Toxicol.*, 26, 39, 1986.

Aldridge, W. N., An assessment of the toxicological properties of pyrethroids and their neurotoxicity, *Crit. Rev. Toxicol.*, 21, 89, 1990.

Aldridge, W. N. and Barnes, J. M., Some problems in assessing the toxicity of the organophosphorus insecticides towards mammals, *Nature (London)*, 169, 345, 1952.

Aldridge, W. N., Barnes, J. M., and Johnson, M. K., Studies on delayed neurotoxicity produced by some organophosphorus compounds, *Ann. N.Y. Acad. Sci.*, 160, 314, 1969.

Aldridge, W. N. and Brown, A. W., The biological properties of methyl and ethyl derivatives of tin and lead, in *The Biological Alkylation of Heavy Metals*, Craig, P. J. and Glockling, F., Eds., Special Publication No. 66, Royal Society of Chemistry, London, 1988, 147.

Aldridge, W. N. and Reiner, E., *Enzyme Inhibitors as Substrates: Interactions of Esterases with Esters of Organophosphorus and Carbamic Acids*, North-Holland, Amsterdam, 1972.

Aldridge, W. N., Street, B. W., and Noltes, J. G. The action of 5-coordinate triorganotin compounds on rat liver mitochondria. *Chem-Biol Interactions*, 34, 223, 1981.

Aldridge, W. N., Street, B. W., and Skilleter, D. N., Halide-dependent and halide-independent effects of triorganotin and triorganolead compounds on mitochondrial functions, *Biochem. J.*, 168, 353, 1977.

Ballantyne, B. and Marrs, T. C., Eds., *Clinical and Experimental Toxicology of Organophosphates and Carbamates*, Butterworth/Heinemann, Oxford, 1992.

Barnes, J. M. and Denz, F. A., Experimental demyelination with organophosphorus compounds, *J. Pathol. Bacteriol.*, 65, 597, 1953.

Barnes, J. M. and Stoner, H. B., Toxic properties of some dialkyl and trialkyltin salts, *Brit. J. Industr. Med.*, 15, 15, 1958.

Barnes, J. M. and Verschoyle, R. D., Toxicity of a new pyrethroid insecticide, *Nature (London)*, 248, 711, 1974.

Bernard, C., *La Science Experimentale*, Bailliere, Paris, 1875.

Bidstrup, P. L., Bonnell, J. A., and Beckett, A. G. Paralysis following poisoning by a new organophosphorus insecticide, *Brit. Med. J.*, 1, 1068, 1953.

Brodie, M. E. and Aldridge, W. N., Elevated cerebellar cyclic GMP levels during deltamethrin-induced motor syndrome, *Neurobehav. Toxicol. Teratol.*, 4, 109, 1982.

Brown, A. W., Aldridge, W. N., Street, B. W., and Verschoyle, R. D. The behavioural and neuropathologic sequelae of intoxication by trimethyltin in the rat, *Amer. J. Path.*, 97, 59, 1979.

Caroldi, S., Magos, L., Jarvis, J., Forshaw, P., and Snowden, R. The potentiation of the non-behavioural effects of amphetamine by carbon disulphide, *J. Appl. Toxicol.*, 7, 63, 1987.

Chambers, J. E., The role of target site activation of phosphorothionates in acute toxicity, in *Organophosphates: Chemistry, Fate and Effects*, Chambers, J. E. and Levi, P. E., Eds., Academic Press, San Diego, 1992, 229.

Clothier, B. and Johnson, M. K., Rapid aging of neurotoxic esterase after inhibition by di-isopropyl phosphorofluoridate, *Biochem. J.*, 177, 549, 1979.

Clothier, B. and Johnson, M. K., Reactivation and aging of neurotoxic esterase inhibited by a variety of organophosphorus esters, *Biochem. J.*, 185, 739, 1980.

Cremer, J. E., The biochemistry of organotin compounds: the conversion of tetraethyltin into triethyltin, *Biochem. J.*, 68, 685, 1958.

Cremer, J. E., Biochemical studies on the toxicity of tetraethyllead and other lead compounds, *Brit. J. Indust. Med.*, 16, 191, 1959.

Cremer, J. E., Amino acid metabolism in rat brain studied with ^{14}C-labelled glucose, *J. Neurochem.*, 11, 165, 1964.

Cremer, J. E., Cunningham, V. J., Ray, D. E., and Sarna, G. S., Regional changes in brain glucose utilization in rats given a pyrethroid insecticide, *Brain Res.*, 194, 278, 1980.

Cremer, J. E., Ray, D. E., Sarna, G. S., and Cunningham, V. J., A study of the kinetic behaviour of glucose based on simultaneous estimation of influx and phosphorylation in brain regions of rats in different physiological states, *Brain Res.*, 221, 331, 1981.

Cunningham, V. J. and Cremer, J. E., A method for the simultaneous estimation of regional rates of glucose influx and phosphorylation using radiolabelled 2-deoxyglucose, *Brain. Res.*, 221, 319, 1981.

Edwards, P. M., Neurotoxicity of acrylamide and its analogues and effect of these analogues, and other agents on acrylamide neuropathy, *Brit. J. Industr. Med.*, 32, 31, 1975.

Eto, M., Casida, J. E., and Eto, T., Hydroxylation and cyclisation reaction involved in the metabolism of tri o-cresylphosphate, *Biochem. Pharmacol.*, 11, 337, 1962.

Forshaw, P. J., Lister, T., and Ray, D. E., Inhibition of a neuronal voltage-dependent chloride channel by the type II pyrethroid, deltamethrin, *Neuropharmacol.*, 32, 105, 1993.

Forshaw, P. J. and Ray, D. E., A novel action of deltamethrin on membrane resistance in mammalian skeletal muscle and non-myelinated nerve fibres, *Neuropharmacol.*, 29, 75, 1990.

Friedberg, T., Siegert, P., Grassow, M. A., Bartlomowicz, B., and Oesch, F., Studies on the expression of the cytochrome P450IA, P450IIB and P450IIC gene family in extrahepatic and hepatic tissues, *Environ. Hlth. Perspec.*, 88, 67, 1990.

Fullerton, P. M. and Barnes, J. M., Peripheral neuropathy in rats produced by acrylamide, *Brit. J. Industr. Med.*, 23, 210, 1966.

Gage, J. C., A cholinesterase inhibitor derived from OO-diethyl O-p-nitrophenyl thiophosphate *in vivo*, *Biochem. J.*, 54, 426, 1953.

Heath, D. F., *Organophosphorus Compounds: Anticholinesterases and Related Compounds*, Pergamon Press, Oxford, 1961.

Holan, G., Rational design of insecticides, *Bull. Wld. Hlth. Org.*, 44, 355, 1971.

Jacobs, J. M., Cremer, J. E., and Cavanagh, J. B., Acute effects of triethyltin on the rat myelin sheath, *Neuropath. Appl. Neurobiol.*, 3, 169, 1977.

Johnson, M. K., The delayed neurotoxic effect of some organophosphorus compounds: identification of the phosphorylation site as an esterase, *Biochem. J.*, 114, 711, 1969.

Johnson, M. K., Organophosphorus and other inhibitors of brain "neurotoxic esterase" and the development of delayed neuropathy in hens, *Biochem. J.*, 120, 523, 1970.

Johnson, M. K., The primary biochemical lesion leading to delayed neurotoxic effects of some organophosphorus esters, *J. Neurochem.*, 23, 785, 1974.

Johnson, M. K., The delayed neuropathy caused by some organophosphorus esters: mechanism and challenge, *Crit. Rev. Toxicol.*, 3, 289, 1975a.

Johnson, M. K., Organophosphorus esters causing delayed neurotoxic effects; mechanism of action and structure-activity studies, *Arch. Toxicol.*, 34, 259, 1975b.

Johnson, M. K., The target for initiation of delayed neurotoxicity by organophosphorus esters: biochemical studies and toxicological applications, *Rev. Biochem. Toxicol.,* 4, 141, 1982.

Johnson, M. K. and Lauwerys, R., Protection by some carbamates against the delayed neurotoxic effects of di-isopropyl phosphorofluoridate, *Nature (London),* 222, 1066, 1969.

Johnson, M. K. and Safi, J. M., Organophosphoramidation of neuropathy target esterase (NTE) is sufficient to initiate delayed neuropathy (DN) without the necessity of an "aging" reaction, *Toxicologist,* 12, 40, Abstract 63, 1992.

Kimbrough, R. D. and Gaines, T. B., Hexachlorophene effects on the rat brain, *Arch. Environ. Hlth.,* 23, 114, 1971.

Koelle, G. B., Ed., *Cholinesterases and Anticholinesterase Agents,* Springer, Berlin, 1963.

Lier, R. B. L. van and Cherry, L. D., The toxicity and mechanism of action of Bromethalin: a new single feeding rodenticide, *Fund. Appl. Toxicol.,* 11, 664, 1988.

Lijinsky, W. and Aldridge, W. N., Increase in cerebral fluid in rats after treatment with triethyltin, *Biochem. Pharmacol.,* 21, 481, 1975.

Lock, E. A., Increase in cerebral fluid in rats after treatment with hexachlorophane or triethyltin, *Biochem. Pharmacol.,* 25, 1455, 1976.

Lock, E. A., The production of an oedematous change in the white matter of the central nervous system of the rat by 3,5-diiodo-4'-chlorosalicylamide, *Proc. Int. Soc. Neurochem.,* 6, 357, 1977.

Lock, E. A. and Berry, L. N., Biochemical changes in the rat cerebellum following cypermethrin administration, *Toxicol. Appl. Pharmacol.,* 59, 508, 1981.

Lock, E. A., Scales, M. D. C., and Little, R. A., Observations on 2'-chloro-2,4-dinitro-5'6-di(trifluoromethyl)-diphenylamine-induced edema in the white matter of the central nervous system of the rat, *Toxicol. Appl. Pharmacol.,* 60, 121, 1981.

Lotti, M., Caroldi, S., Capodicasa, E. and Moretto, A., Promotion of organophosphorus-induced delayed polyneuropathy by phenylmethanesulfonyl fluoride, *Toxicol. Appl. Pharmacol.,* 108, 234, 1991.

Magee, P. N., Stoner, H. B., and Barnes, J. M., The experimental production of oedema in the central nervous system of the rat by triethyltin compounds, *J. Pathol. Bacteriol.,* 53, 107, 1957.

Magos, L., Neurotoxicity, anorexia and the preferential choice of antidote in methylmercury intoxicated rats, *Neurobehav. Toxicol Teratol.,* 4, 643, 1982.

Magos, L., Brown, A. W., Sparrow, S., Bailey, E., Snowden, R. T., and Skipp, W. R., The comparative toxicology of ethyl- and methyl-mercury, *Arch. Toxicol.,* 57, 260, 1985.

Magos, L. and Butler, W. H., The kinetics of methylmercury administered repeatedly to rats, *Arch. Toxicol.,* 35, 25, 1976.

Minn, A., Ghersa-Egea, J. F., Perrin, R., Leininger, B., and Siest, G., Drug metabolising enzymes in the brain and cerebral microvessels, *Brain Res. Rev.,* 16, 65, 1991.

Moretto, A., Capodicasa, E. and Lotti, M., Clinical expressions of organophosphate-induced delayed polyneuropathy in rats, *Toxicol. Lett.,* 63, 97, 1992.

Moretto, A., Capodicasa, E., Periaca, M. and Lotti, M., Age sensitivity to organophosphorus-induced delayed polyneuropathy: biochemical and toxicological studies in developing chicks, *Biochem. Pharmacol.,* 41, 1497, 1991.

Narahashi, T., Chemicals as tools in the study of excitable membranes, *Physiol. Rev.,* 54, 813, 1974.

O'Brien, R. D., *Toxic Phosphorus Esters; Chemistry, Metabolism and Biological Effects,* Academic Press, New York, 1960.

Peters, R. A., *Biochemical Lesions and Lethal Synthesis,* Pergamon Press, Oxford, 1963.

Peters, R. A., The biochemical lesion and its historical development, in *Mechanisms of Toxicity,* W. N. Aldridge, Ed., *Brit. Med. Bull.,* 25, 223, 1969.

Pope, C. N. and Padilla, S., Potentiation of organophosphorus-induced delayed neurotoxicity by phenylmethanesulfonyl fluoride, *Toxicol. Environ. Hlth.,* 31, 261, 1990.

Ray, D. E., Brown, A. W., Cavanagh, J. B., Nolan, C. C., Richards, H. K., and Wyllie, S. P., Functional/metabolic modulation of the brain stem lesions caused by 1,3-dinitrobenzene in the rat, *Neurotoxicol.*, 13, 379, 1992.

Rickard, J. and Brodie, M. E., Correlation of blood and brain levels of the neurotoxic pyrethroid deltamethrin with the onset of symptoms in rats, *Pestic. Biochem. Physiol.*, 23, 143, 1985.

Romero, I., Brown, A. W., Cavanagh, J. B., Nolan, C. C., Ray, D. E., and Seville, M. P., Vascular factors in the neurotoxic damage caused by 1,3-dinitrobenzene in the rat, *Neuropath. Appl. Neurobiol.*, 17, 495, 1991.

Saunders, B. C., *Some Aspects of the Chemistry and Toxic Action of Organic Compounds Containing Phosphorus and Fluorine*, University Press, Cambridge, 1957.

Seawright, A. A., Brown, A. W., Ng, J. C., and Hrdlicka, J., Experimental pathology of short-chain alkyllead compounds, in *Biological Effects of Organolead Compounds,* Grandjean, P. and Grandjean, E. C., Eds., CRC Press, Boca Raton, 1984, 177.

Selwyn, M. J., Triorganotin compounds as ionophores and inhibitors of translocating ATPases, in *Organotin Compounds: New Chemistry and Applications*, Zuckerman, J. J., Ed., Adv. Chem. Series, 157, 204-226, American Chemical Society, Washington, 1976.

Smith, M. I., Elvove, E., and Frazier, W. H., The pharmacological action of certain phenol esters with reference to the relation of chemical and physiologic action, *U.S. Public Hlth. Rep.*, 45, 2509, 1930.

Toggas, S. M., Krady, J. K., and Billingsley, M. L., Molecular neurotoxicology of trimethyltin: identification of stannin, a novel protein expressed in trimethyltin-sensitive cells, *Mol. Pharmacol.*, 42, 44, 1992.

van der Kerk, G. J. M., Organotin chemistry: past, present and future, in *Organotin Compounds: New Chemistry and Applications*, Adv. Chem. Series, 157, 1–24, American Chemical Society, Washington, 1976.

Verschoyle, R. D. and Aldridge, W. N., Structure-activity relationships of some pyrethroids in rats, *Arch. Toxicol.,* 45, 325, 1980.

Verschoyle, R. D., Brown, A. W., and Thompson, C. A., The toxicity and neuropathology of 2,4,5-tribromimidazole and its derivatives in the rat, *Arch. Toxicol.*, 56, 109, 1984.

Vijverberg, H. P. M. and van den Bercken, J., Action of pyrethroid insecticides on the vertebrate nervous system, *Neuropath. Appl. Neurobiol.*, 8, 421, 1982.

Vijverberg, H. P. M. and van den Bercken, J., Neurotoxicological effects and the mode of action of pyrethroid insecticides, *Crit. Rev. Toxicol.,* 21, 105, 1990.

World Health Organization, Alternative insecticides for vector control, *Bull. Wld. Hlth. Org.,* 44, 1, 1971.

Yoshimoto, T., Ogawa, S., Udagawa, T., and Numata, S., Development of a new insecticide, etofenprox, *J. Pestic. Sci.,* 14, 259, 1989.

Chapter 2

INJURY AND REPAIR IN NEUROTOXICOLOGY

Marcello Lotti

CONTENTS

I. Introduction 19

II. Neurotoxicity of Pyrethroids 21

III. Neurotoxicity of MPTP and Neurodegenerative Disorders 23

IV. Delayed Polyneuropathy Caused by Organophosphates and
Other Esterase Inhibitors 26

V. Conclusions 30

Acknowledgments 30

References 31

I. INTRODUCTION

Perturbation of normal life processes by toxic chemicals enables us to learn about the life processes themselves (Bernard, 1875). This concept, though more than 100 years old, is still valid and represents an important aspect of toxicology. Moreover, life processes are highly dynamic because of the continuous turnover of cells, subcellular structures, and biological molecules, which is based upon death-regeneration, damage-repair, injury-compensation, and degradation-resynthesis. Therefore, in order to define toxicity more precisely and to broaden the approach to mechanistic studies, the result of exposures to chemicals might be regarded as the product of divergent biological processes: injury on the one hand and compensation/repair on the other hand. The key concepts of dose-response and thresholds (the latter encountered both in toxicology and in physiology) indirectly suggest that, before a given biological effect is expressed, processes do exist which are able to handle biochemical and physiological changes induced by chemicals, or by life itself, without great expenditure. Therefore, exposures to chemicals may result in toxicity either

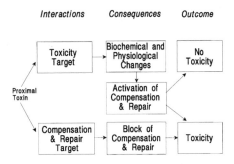

FIGURE 2.1. Diagram representation of consequences and outcome of chemical/target interactions.

because injury overwhelms compensation and repair or because chemicals interfere directly with compensation/repair systems, or both.

Figure 2.1 illustrates these concepts assuming that chemicals cause biological effects through specific interactions with molecular targets (Aldridge, 1986). The pathophysiological consequences of interactions with toxicity target(s) activate mechanisms of compensation and repair which in turn will or will not be effective in counteracting the processes leading to toxicity. Chemicals might also directly affect compensation/repair targets, which are involved in the responses to either physiologically or environmentally induced damage. In such cases, toxicity would either be due to the block of physiological turnover, thereby causing the accumulation of physiological deficit, or to the amplification of a concurrent injury, respectively. Keeping in mind Claude Bernard's dictum, this paper discusses some examples of neurotoxicity using this perspective.

It is possible to learn a great deal about how the nervous system works simply by paying attention to four general features (Kandel, 1985):

1. The mechanism by which nerve cells produce their signals
2. The ways in which nerve cells are connected
3. The relationship of various patterns of interconnections to different types of behavior
4. The means by which nerve cells and their connections are modified

Explanations for some of these features of the nervous system are available, and the use of neurotoxins has been shown to be pivotal to this understanding (Price and Griffin, 1980). By using neurotoxic chemicals, will it now be possible to contribute further to our knowledge of these general features by analyzing and separating the primary actions of chemicals on toxicity targets from the effects on compensation/repair mechanisms and targets?

As examples for this approach, I will discuss three neurotoxicants: pyrethroid insecticides, the meperidine-like drug MPTP (1,-methyl-4-phenyl-1,2,3,6-

tetrahydropyridine) and, finally, organophosphates (OPs) together with certain other esterase inhibitors.

II. NEUROTOXICITY OF PYRETHROIDS

The general flow of information both within and between the cells of the nervous system is conveyed by transient electrical signals. The basic mechanism is a change of the electrical properties of the cell membrane generating action potentials which are essential for the long-distance transmission between hierarchically connected neurons. Action potentials through voltage-gated signals are one of these electrical signals (Koester, 1985), and they can be investigated by means of neurotoxic chemicals, such as pyrethroids, known to act at voltage-dependent sodium channels (Vijverberg and van den Bercken, 1990).

The generation of action potentials through voltage-gated channels is an all or none phenomenon which is mediated by Na^+ influx through sodium channels. This threshold characteristic is due to the activation, during depolarization, of K^+ channels thereby opposing, up to a certain point, the depolarizing action of Na^+ influx. This activation of K^+ channels represents a physiological compensation mechanism which might influence a physiological or toxic response of Na^+ channels. Considering that the pattern and distribution of voltage-sensitive channels varies among neurons and that in a given neuron's cell body it determines how that cell will respond to synaptic input, then an effect on such channels will result in selective toxicity for some cells. In the case of pyrethroid insecticides, the behavioral changes induced (Verschoyle and Aldridge, 1980) might be the result of the marked diversity of input-output characteristics displayed by different neurons, which is in turn due to cell-to-cell variations in the combination of channel types that differ in ion selectivity, activation kinetics, and voltage sensitivity.

Pyrethroid insecticides have been divided into two classes according to their effects in mammals and insects, which depend on the chemical structure, i.e., on the presence of an α-cyano group in the molecule. Thus, noncyano pyrethroids cause in rats the so-called T-syndrome, characterized by tremors; while α-cyano pyrethroids cause the CS-syndrome, consisting of salivation, writhing (choreoathetosis), and seizures. For a few pyrethroids the two syndromes overlap, but in general this distinction is clear (Verschoyle and Aldridge, 1980; Aldridge, 1990). Pyrethroids reversibly interact with macromolecular component(s) of the voltage-dependent sodium channel, delaying the closing of the sodium channel activation gate, thereby resulting in a prolongation of the inward sodium current during excitation. A given time constant is characteristic for each pyrethroid, and in general there is a correlation between time constants and the two clinical syndromes caused in rats (Vijverberg and van den Bercken, 1990). Cyano pyrethroids have a more prolonged effect on the sodium channel than noncyano pyrethroids. An example of such differences is given in Table 2.1 where *in vivo* data are also reported. The two insecticides,

TABLE 2.1
Comparative Toxicology of Noncyano (Cismethrin) and Cyano (Deltamethrin) Pyrethroids: *In Vitro* Electrophysiological Effects, Acute Toxicity, and Treatment

	Cismethrin	Deltamethrin
Time constants of sodium tail currents (ms)[a,b]	28	1770
Syndrome in rats[c,b]	T	CS
Lethal dose [μmol · kg^{-1}(i.v.)][d,b]	30	10
Prevention of motor symptoms and lethality by mephenesin infusion [μmol · kg^{-1} · min^{-1} (i.v.)][d]	55	22

[a]   From Vijverberg and van den Bercken, 1990.
[b]   Similar results have been reported by Wright et al., 1988.
[c]   From Verschoyle and Aldridge, 1980.
[d]   From Bradbury et al., 1983.

cismethrin and deltamethrin, display a different pattern of motor symptoms when given intravenously at lethal doses. Comparisons of brain concentrations of both chemicals obtained after lethal doses indicate that there are no major differences in their delivery to the target organ and that their dissociation from the target should also not be very different, given similar reduction of effects during the recovery from acute symptoms (Aldridge, 1990). Therefore the effects of cismethrin and deltamethrin seem to be dependent on their intrinsic activity, once they have become attached to the macromolecules of sodium channels. Table 2.1 also shows that a centrally active muscle relaxant, mephenesin, is able to abolish motor symptoms and mortality caused by both chemicals. The therapeutic action of mephenesin is thought to be due to the depression of polysynaptic pathways in the spinal cord (Bradbury et al., 1981; 1983). Suggestions were made for a possible common mechanism between mephenesin and other substances which are also effective in the treatment of pyrethroid toxicity (Aldridge, 1990) (Figure 2.2). While deltamethrin has a more sustained effect on sodium channels than cismethrin, the motor symptoms caused by deltamethrin are abolished by doses of mephenesin smaller than those required to have the same effect on cismethrin toxicity. It can be concluded that the two pyrethroids act on different neurons (Wright et al., 1988) and/or that compensation for the toxic effects of the two pyrethroids is different and perhaps involves more than one mechanism. Moreover, since deltamethrin inhibits voltage-gated chloride channels, CS-pyrethroid toxicity might also result by preventing the compensation function of these latter channels which stabilize membrane potentials (Forshaw et al., 1993; Forshaw and Ray, 1993). The identification of the mechanism of action of these drugs in ameliorating the toxicity of pyrethroids will enable a better understanding of the relationships among tail currents, toxicity and compensation, and perhaps of the physiological relevance of sodium channels in various parts of the nervous system.

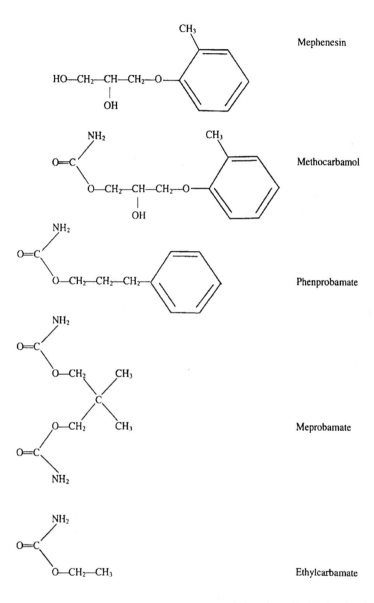

FIGURE 2.2. Structure of drugs capable of reducing lethality of pyrethroids. Reprinted with permission from Aldridge, W.N., An assessment of the toxicological properties of pyrethroids and their neurotoxicity, *Critical Reviews in Toxicology,* 21, 89, 1990. Copyright CRC Press, Boca Raton, Florida.

III. NEUROTOXICITY OF MPTP AND NEURODEGENERATIVE DISORDERS

The cellular organization of the brain is complex, and it has been subdivided into several patterns of neuronal connectivity (Bloom, 1990). One of these is utilized by certain neuronal systems of the hypothalamus, pons, and brainstem. From a single anatomical location these neurons extend multiple-branched and divergent connections to many target cells outside the brain region in which the neurons are located, indicating a larger spatial domain. These systems might contain neurotransmitters, neuropeptides, or hormones, and their purpose is to mediate linkages between regions that require integration. Integration is maintained even if the system is partially and selectively destroyed, such as in the case of lesions of the nigrostriatal tract leading to Parkinson's disease.

Parkinson's disease is due to the destruction of the pigmented nigral neurons of the brain. It is believed that for the appearance of clinical symptoms of Parkinson's disease a long period is required, when nigral degeneration progresses. In fact, in people dying with overt parkinsonism a substantial loss of nigral neurons (50% or more) and striatal dopamine depletion (80% or more) are found, suggesting that several middle-aged and elderly individuals with less than this loss of neurons and dopamine depletion might have preclinical Parkinson's disease; these individuals, if they live long enough, may develop the disease in later life (Marsden, 1990). Methods to identify such people are being developed, among which positron emission tomography (PET) (Perlmutter, 1988). With this technique it is possible to measure after intravenous injection of [^{18}F]6-fluorodopa, its accumulation in human striatum, which reflects the striatal storage capacity for dopamine. In this way, it is possible to detect the extent of nigrostriatal degeneration (Leenders et al., 1986) and, indirectly, to titrate the compensation capability of the entire system.

The research on Parkinson's disease has progressed astonishingly over the last 30 years, and the discovery of MPTP-induced parkinsonism represents a cornerstone (Davis et al., 1979; Langston et al., 1983). MPTP is an impurity arising as a by-product of meperidine-analog synthesis which selectively destroys the substantia nigra, thereby inducing neuropathological, neurochemical, and clinical changes similar to Parkinson's disease. PET studies in asymptomatic subjects who were exposed to MPTP showed that they had a damaged nigrostriatal pathway and led to the hypothesis that the toxic effects of MPTP will appear later because they will be added to the physiological cell loss occurring as the normal consequence of aging (Calne et al., 1985).

MPTP represents, therefore, a useful tool to address several important questions in terms of compensation mechanisms, for instance: How is the up to 80% loss of dopamine compensated for? Why is the inhibitory action of dopamine fibers on the striatal cholinergic neurons maintained until a 50% loss of these neurons is seen? It might take a long time before we have answers, but the understanding of the toxic action of MPTP and the studies on the pathogenesis of neurodegenerative disorders have already highlighted some aspects of the compensation/repair mechanisms. The biochemical mechanism of the selective neurotoxicity of MPTP to nigral dopamine-containing cells has been extensively investigated (Singer et al., 1987) (Figure 2.3). In short, MPTP is activated by MAO-B to MPP$^+$ (1-methyl-4-phenylpyridinium) which enters the cell through the dopamine reuptake system and into mitochondria through an energy-dependent carrier. The toxic action of MPP+ is thought to be due to the inhibition of NADH dehydrogenase, leading to cessation of oxidative phosphorylation, depletion of ATP supply, and cytotoxicity. However, it has also been shown that MPP+ toxicity is prevented by N-methyl-D-aspartate (NMDA) antagonists (Turski et al., 1991) (Figure 2.4), suggesting that MPP+ toxicity may be mediated by excitatory amino acids. A possible mechanism would be mediated by energy depletion, changes in cell membrane, and by the relief of the block of NMDA receptor channels, enabling glutamate to become neurotoxic.

Excitatory amino acids are thought to be involved in a number of neurodegenerative diseases, including parkinsonism (Choi, 1988). The hypothesis of a common mechanism for neurodegenerative disorders is based on the clinical characteristics of Guam disease, the amyotrophic lateral sclerosis (ALS) parkinsonism and dementia complex, which was related with the ingestion of excitatory amino acids (Spencer et al., 1987). Even though doubts on the etiology of Guam disease are increasing (Travis, 1993), there is still substantial evidence of a contribution of excitotoxicity to neuronal loss in most neurodegenerative disorders.

For instance, the involvement of excitotoxicity in ALS is shown by the deficit of a carrier system responsible for the clearance of glutamate from the

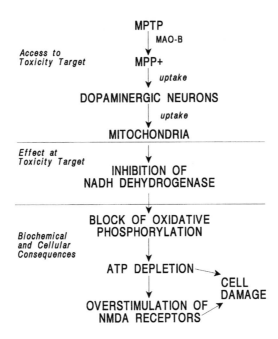

FIGURE 2.3. Mechanisms of MPTP (1,-methyl-4-phenyl-1,2,3,6-tetrahydropyridine) toxicity to nigrostriatal cells.

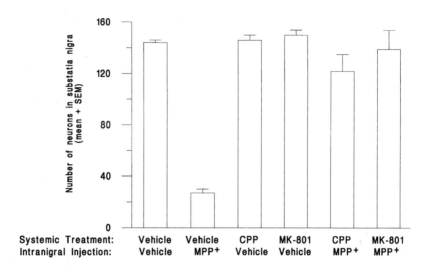

FIGURE 2.4.   Protection from MPP+ (1-methyl-4-phenylpyridinium) neurotoxicity in rats by two N-methyl-D-aspartate antagonists (data from Turski et al., 1991).

Systemic treatments with vehicle, CPP (3-((±)-2-carboxypiperazine-4-yl)-propyl-1-phosphonate) (0.1 mmol · kg^{-1} × 18) and MK-801 ((+)-5-methyl-10,11-dihydro-5H-dibenzo-(a,d) cycloheptan-5,10-imine maleate (0.01 mmol · kg^{-1} × 18) were given every 4 hours starting 30 min before MPP+. MPP+ (0.25 μmol) was injected into one substantia nigra. The contralateral was injected with vehicle as a control. Morphometric analysis was performed 7 days after MPP+ injection.

synaptic cleft, which was found to be associated with sporadic cases of ALS (Rothstein et al., 1992). This observation points to the relevance of abnormal cellular metabolism of glutamate, thereby suggesting that excitotoxicity might be the result of the impairment of the mechanisms which remove the excess of the excitatory amino acid. This physiological function might be regarded as a mechanism which protects from otherwise toxic excess of glutamate. Indeed, other compensation/repair mechanisms might be involved in the pathogenesis of ALS, such as the inhibition of axonal sprouting (Gurney et al., 1984), altering in this way neuronal plasticity. These data indicate that influences on compensation/repair might have critical relevance in the pathogenesis of certain neurological disorders. Considering that, as indicated above, also MPTP toxicity may be mediated by excitotoxicity, it might be that a certain degree of protection from excitotoxicity accounts for the compensation/repair phenomena which are effective when the dopaminergic cells are already reduced, but Parkinson's disease is not yet clinically detectable.

IV. DELAYED POLYNEUROPATHY BY ORGANOPHOSPHATES AND OTHER ESTERASE INHIBITORS

The nervous system is highly plastic because nerve cells and their connections are continuously modified. Several factors including hormonal status, environmental influence, and injury mediate such adaptive responses both in the central and peripheral nervous system. Axonal sprouting and concurrent synaptic turnover are the ongoing plastic processes responsible for adaptation responses, and they are known to occur also in the absence of injury (Cotman and Anderson, 1989).

In the peripheral nervous system, axonal sprouting can restore damaged circuits and participate in functional recovery. The regeneration of peripheral axons after injury requires the presence of their supporting Schwann cells and probably also of extraneuronal cells. The requirement for the Schwann cell is specific because, when regenerating axons are exposed to central glial cells, such as astrocytes and oligodendrocytes, their normal regenerating outgrowth is inhibited (Kelly, 1985). Studies in certain mutant mice have also shown that unknown nerve factors are able to recruit monocytes which play a role in activating Wallerian degeneration following peripheral nerve transection (Perry et al., 1990).

It is possible that certain esterase inhibitors would influence these or other mechanisms, since central-peripheral axonopathies of different origin were all found to be exacerbated by these chemicals, which hardly cause neurotoxicity by themselves (Lotti and Moretto, 1993).

Some organophosphorus esters (OP) cause a central-peripheral degeneration of distal axons known as organophosphate-induced delayed polyneuropathy (OPIDP) (Johnson, 1982, 1990; Lotti, 1992). Phosphates, phosphoroamidates, and phosphonates cause OPIDP when they inhibit, soon after dosing, more than 70% of neuropathy target esterase (NTE), a protein of nervous tissue thought to be the target of OPIDP. Other NTE inhibitors such as phosphinates, carbamates, and sulfonyl halides cause either protection from OPIDP (Johnson and Lauwerys, 1969) when given before, or promotion of OPIDP (Pope and Padilla, 1990; Lotti et al., 1991) when given after a neuropathic OP. The protective effect is related to the inhibition of NTE (at least 30 to 40%) that is caused prior to dosing with a neuropathic OP (Johnson and Lauwerys, 1969). However, on the contrary, it seems that promotion is not related to NTE inhibition, even though in most cases the dose promoting neuropathy was also inhibitory of NTE (Lotti and Moretto, 1993). Furthermore when these protective-promoting compounds are given at very high and repeated doses, they cause a mild neuropathy similar to OPIDP, and in this case an almost complete inhibition of NTE is observed (Lotti et al., 1993).

The physiological functions of NTE are, at present, unknown. It is difficult to maintain that NTE catalytic activity is essential to the normal function of neurons, given the substantial inhibition of NTE (for instance 80%) observed in the absence of any neuropathy, after dosing with phosphinates, carbamates, and sulfonyl halides. The hypothesis was discussed that NTE might be a receptor of unknown functions where various inhibitors can bind with different "potency," so that different levels of NTE inhibition are necessary in order to trigger neuropathies of comparable severity (Lotti et al., 1993; Lotti and Moretto, 1993).

TABLE 2.2

Promotion of Polyneuropathy by Phenylmethanesulfonyl Fluoride (PMSF) and the Initiating Levels of NTE Inhibition[a]

Compound	Dose (mg/kg/route)	% NTE Inhibition Before PMSF			Maximal Clinical Score[b]	
		Brain	Spinal Cord	Peripheral Nerve	Without PMSF	With PMSF[c]
Diisopropylfluoro-phosphate (DFP)	0.3 s.c.	44,51	58,54	49,42	0(0–0)	2(0–4)
Methamidophos[d]	120 p.o.	76 ± 10	72 ± 10	76 ± 11	0(0–0)	1(1–2)
Dichlorvos[d]	100 s.c.	71 ± 8	44 ± 15	73 ± 10	0(0–0)	1(0–7)
Phenyl N-methyl N-benzyl carbamate (PMBC)[e]	40 i.v.	90,84	ND	ND,80	0(0–0)	2(2–2)

Note: ND = not done.

[a]   From Lotti et al., *Toxicology and Applied Pharmacology*, 122, 165, 1993. With permission.
[b]   Expressed as median (range) on a 0–8 point scale.
[c]   PMSF (120 mg/kg s.c.) given 24 hours after the initiator.
[d]   They cause OPIDP when the level of NTE inhibition is >85% (Johnson et al., 1991; Caroldi and Lotti, 1981).
[e]   It causes neuropathy when the level of NTE inhibition is >95% (Lotti et al., 1993).

Why does OPIDP not develop when less than 70% NTE is affected? Two explanations are possible: no pathophysiological changes are triggered at all by subthreshold inhibitions of NTE because the effect has no adverse consequences, or subthreshold inhibition of NTE represents a biochemical lesion which is efficiently and quickly repaired by some unknown mechanism(s). The phenomenon of promotion allows consideration of the second hypothesis as the more likely. In fact, doses of phenylmethanesulfonyl fluoride, much lower than the neuropathic ones, promote OPIDP which was initiated by diisopropylfluoro-phosphate doses causing about 50% NTE inhibition (Lotti et al., 1991) (Table 2.2). This suggests that a neuropathic biochemical lesion had occurred which could only be expressed by promoters. Recent experiments aimed at understanding OPIDP promotion have demonstrated that the phenomenon is not due to additive effects (Pope et al., 1993) and it can be triggered without inhibition of NTE (Moretto et al., 1994). In fact, promotion cannot be explained as the result of pharmacokinetic interactions because the degree of NTE inhibition caused by a neuropathic OP (expressed as either the duration of inhibition or the maximal inhibition which was achieved) was not affected when a promoter which does not inhibit NTE was also given. In conclusion, it seems that promotion affects a target other than NTE catalytic activity.

The observation that different non-neuropathic levels of NTE inhibition obtained with various OPs are promoted to clinical neuropathy of similar severity by the same dose of promoter suggests that initiation is of variable efficacy in terms of NTE inhibition and depends on the initiating chemical (Table 2.2). Therefore, since promotion affects a target other than that of OPIDP initiation (NTE catalytic activity), it is possible that this target is involved with some compensation or repair event. By blocking it, the severity of clinical outcome will reflect the initiating biochemical lesion only.

This mechanism(s) of repair may well be more efficient at a very young age (Johnson and Barnes, 1970; Moretto et al., 1991). Table 2.3 shows that the sensitivity to OPIDP develops with age both in hens and rats and is not related to the levels of NTE inhibition. In fact, NTE is equally and almost entirely inhibited at all ages and yet only older animals develop OPIDP. These results suggest that the efficacy of compensation/repair mechanisms declines with age. Promotion experiments in chicks further indicate, as in the hen, that once promotional mechanism is activated (possibly involving a block of the repair process), then the severity of OPIDP depends on the initiating OP only (Peraica et al., 1993).

There is no evidence that the mechanism involved in the promotion of OPIDP, assuming the block of compensation/repair, is the same mechanism which compensates for the lower than threshold inhibitions of NTE.

Promotion was also found to occur when the promoter was given at doses which do not inhibit NTE before the neuropathic insult (Moretto et al., 1994), suggesting that it does not interfere with the cascade of events triggered by the initiating biochemical lesion. Animals, therefore, seem to be "primed" in some way by the promoter, committing themselves to an exaggerated toxic response. It is possible that this pre-existing compensation mechanism will also function in the absence of external injury, thereby representing a mechanism of nervous system plasticity involved in the physiological turnover of long axons. Conse-

TABLE 2.3

Age-Related Sensitivity to Organophosphate Polyneuropathy (OPIDP) of Hens and Rats[a]

	Hen			Rat		
Age (days)	40	60	>100	30	110	>180
% NTE inhibition[b]	84	80	88	92	ND	95
OPIDP[c]	−[d]	±	+	−[d]	±	+

Note: ND = not done.

[a] Data from Moretto et al., 1991, 1992a and Peraica et al., 1993.
[b] In brain, 24 hours after dosing hens and rats with DFP (1 mg/kg s.c.) and DBDCVP (5 mg/kg s.c.), respectively.
[c] Clinical assessment: − = no neuropathy; ± = marginal neuropathy; + = neuropathy.
[d] Promotable to OPIDP when PMSF follows.

quently, if repeated small doses of promoters cause neuropathy, this might be due to the expression of accumulated physiological damage.

Promotion was also found to occur in neuropathies other than OPIDP, such as that induced by 2,5-hexanedione and by nerve crush (Moretto et al., 1992b; 1993), known to have a different pathogenesis. It seems therefore that this unknown repair mechanism which is thought to be affected by promotion is effective in other types of axonopathy, too.

V. CONCLUSIONS

Recent research in neurobiology and neurotoxicology suggests that compensation/repair mechanisms of the nervous system can be explored. In this way the studies of nervous system toxicology become broader, indicating that impairment of compensation/repair might lead to permissive progression of injuries. The identification of molecular targets of toxicity, such as the sodium channels for pyrethroid toxicity, the mitochondrial NADH dehydrogenase of nigrostriatal cells for MPTP, and the NTE for OPIDP allows a quantitative assessment of toxic effects at the molecular level. The evidence that certain chemicals modulate toxic responses following toxin-target interactions, such as muscle relaxants for pyrethroid toxicity, excitotoxic amino acids for MPTP toxicity, and esterase inhibitors for axonopathies, provides tools to assess the relevance and the extent of compensation/repair mechanisms. Finally, when the targets and mechanisms for these modulatory effects are identified, they will shed new light on the physiology of the nervous system.

The expansion of toxicity studies into the mechanisms of compensation/repair of the nervous system might also lead to practical exploitations and most important, to the understanding of the dynamics of nervous system physiological processes which will open the way to rational therapeutic strategies for neurological diseases.

ACKNOWLEDGMENTS

I dedicate this paper to Martin K. Johnson, a pioneer in neurotoxicology, a generous mentor, and a good friend, on the occasion of his retirement from the MRC Toxicology Unit. I thank W.N. Aldridge, A. Moretto, and D. Ray for their valuable comments and C.A. Drace-Valentini for manuscript preparation. The financial support of CNR, Ministero Italiano dell'Universita' e della Ricerca Scientifica e Tecnologica, and Regione Veneto is gratefully acknowledged.

REFERENCES

Aldridge, W. N., The biological basis and measurement of thresholds, *Ann. Rev. Pharmacol. Toxicol.,* 26, 39, 1986.

Aldridge, W. N., An assessment of the toxicological properties of pyrethroids and their neurotoxicity, *Crit. Rev. Toxicol.,* 21, 89, 1990.

Bernard, C., *La Science Expérimentale,* Ballière, Paris, 1875.

Bloom, F. E., Neurohumoral transmission and the central nervous system, in Goodman and Gilman's *The Pharmacological Basis of Therapeutics,* 8th edition, Goodman Gilman, A., Rall, T. W., Nies, A. S., and Taylor, P., Eds., Pergamon Press, New York, 1990, 244.

Bradbury, J. E., Forshaw, P. J., Gray, A. J., and Ray D. E., The action of mephenesin and other agents on the effects produced by two neurotoxic pyrethroids in the intact and spinal rat, *Neuropharmacology,* 22, 907, 1983.

Bradbury, J. E., Gray, A. J., and Forshaw, P., Protection against pyrethroid toxicity in rats with mephenesin, *Toxicol. Appl. Pharmacol.,* 60, 382, 1981.

Calne, D. B., Langston, J. W., Martin, W. R. W., Stoessl, A. J., Ruth, T. J., Adam, M. K., Pate, B. D., and Schulzer, M., Positron emission tomography after MPTP: observations relating to the cause of Parkinson's disease, *Nature,* 317, 246, 1985.

Caroldi, S. and Lotti, M., Delayed neurotoxicity caused by a single massive dose of dichlorvos to adult hens, *Toxicol. Lett.,* 9, 157, 1981.

Choi, D. W., Glutamate neurotoxicity and diseases of the nervous system, *Neuron,* 1, 623, 1988.

Cotman, C. W. and Anderson, K. J., Neural plasticity and regeneration, in *Basic Neurochemistry,* 4th edition, Siegel, G., Agranoff, B., Albers, R. W., and Molinoff, P., Eds., Raven Press, New York, 1989, 507.

Davis, G. C., Williams, A. C., Markey, S. P., Herbert, M. H., Caine, E. D., Reichert, C. M., and Kopin, I. J., Chronic parkinsonism secondary to intravenous injection of meperidine analogues, *Psychiatry Res.,* 1, 649, 1979.

Forshaw, P. J., Lister, T., and Ray, D. E., Inhibition of a neuronal voltage-dependent chloride channel by the type II pyrethroid, deltamethrin, *Neuropharmacology,* 32, 105, 1993.

Forshaw, P. J. and Ray, D. E., A voltage-dependent chloride channel in NIE115 neuroblastoma cells is inactivated by protein kinase C and also by the pyrethroid deltamethrin, *J. Physiol.,* 464, 241, 1993.

Gurney, M. E., Belton, A. C. Cashman, N., and Antel, J. P., Inhibition of terminal axonal sprouting by serum from patients with amyotrophic lateral sclerosis, *N. Engl. J. Med.,* 311, 933, 1984.

Johnson, M. K., The target for initiation of delayed neurotoxicity by organophosphorus esters: Biochemical studies and toxicological applications, *Rev. Biochem. Toxicol.,* 4, 141, 1982.

Johnson, M. K., Organophosphates and delayed neuropathy — Is NTE alive and well?, *Toxicol. Appl. Pharmacol.,* 102, 385, 1990.

Johnson, M. K. and Barnes, J. M., Age and the sensitivity of chicks to the delayed neurotoxic effects on some organophosphorus compounds, *Biochem. Pharmacol.,* 19, 3045, 1970.

Johnson, M. K. and Lauwerys, R., Protection by some carbamates against the delayed neurotoxic effects of diisopropyl phosphorofluoridate, *Nature,* 222, 1066, 1969.

Johnson, M. K., Vilanova, E., and Reed, D. J., Anomalous biochemical responses in tests of the delayed neuropathic potential of methamidophos (O,S-dimethyl phosphothioamidate) its resolved isomers and of some higher O-alkyl homologues, *Arch. Toxicol.,* 65, 618, 1991.

Kandel, E. R., Nerve cells and behaviour, in *Principles of Neural Science,* 2nd edition, Kandel, E. R. and Schwartz, J. M., Eds., Elsevier, New York, 1985, 13.

Kelly, J. P., Reactions of neurons to injuries, in *Principles of Neural Science,* 2nd edition, Kandel, E. R. and Schwartz, J. M., Eds., Elsevier, New York, 1985, 187.

Koester, J., Voltage-gated channels and the generation of the action potential, in *Principles of Neural Science,* 2nd edition, Kandel, E. R. and Schwartz, J. M., Eds., Elsevier, New York, 1985, 75.

Langston, J. W., Ballard, P., Tetrud, J. W., and Irwin, I., Chronic parkinsonism in humans due to a product of meperidine-analog synthesis, *Science*, 219, 979, 1983.

Leenders, K. L., Palmer, A. J., Quinn, N., Clark, J. C., Firnau, G., Garnett, E. S., Nahmias, C., Jones, T., and Marsden, C. D., Brain dopamine metabolism in patients with Parkinson's disease measured with positron emission tomography, *J. Neurol. Neurosurg. Psychiatry*, 49, 853, 1986.

Lotti, M., The pathogenesis of organophosphate induced delayed polyneuropathy, *Crit. Rev. Toxicol.*, 21, 465, 1992.

Lotti, M., Caroldi, S., Capodicasa, E., and Moretto, A., Promotion of organophosphate induced delayed polyneuropathy by phenylmethanesulfonyl fluoride, *Toxicol. Appl. Pharmacol.*, 108, 234, 1991.

Lotti, M. and Moretto, A., The search for physiological functions for NTE: is NTE a receptor?, *Chem-Biol. Interact.*, 87, 407, 1993.

Lotti, M., Moretto, A., Capodicasa, E., Bertolazzi, M., Peraica, M., and Scapellato, M. L., Interactions between neuropathy target esterase and its inhibitors and the development of polyneuropathy, *Toxicol. Appl. Pharmacol.*, 122, 165, 1993.

Marsden, C. D., Parkinson's disease, *Lancet*, 335, 948, 1990.

Moretto, A., Bertolazzi, M., Capodicasa, E., Peraica, M., Richardson, R. J., Scapellato, M. L., and Lotti, M., Phenylmethanesulfonyl fluoride elicits and intensifies the clinical expression of neuropathic insults, *Arch. Toxicol.*, 66, 67, 1992b.

Moretto, A., Bertolazzi, M., and Lotti, M., The phosphorothioic acid -O-(2-Chloro-2,3,3-Trifluorocyclobutyl)-O-ethyl S-propyl ester exacerbates organophosphate polyneuropathy without inhibition of neuropathy target esterase, *Toxicol. Appl. Pharmacol.*, 129, 133, 1994.

Moretto, A., Capodicasa, E., and Lotti, M., Clinical expression of organophosphate-induced delayed polyneuropathy in rats, *Toxicol. Lett.*, 63, 97, 1992a.

Moretto, A., Capodicasa, E., Peraica, M., and Lotti, M., Age sensitivity to organophosphate-induced delayed polyneuropathy. Biochemical and toxicological studies in developing chicks, *Biochem. Pharmacol.*, 41, 1497, 1991.

Moretto, A., Capodicasa, E., Peraica, M., and Lotti, M., Phenylmethanesulfonyl fluoride delays recovery from crush of peripheral nerves in hens, *Chem.-Biol. Interact.*, 87, 457, 1993.

Peraica, M., Capodicasa, E., Moretto, A., and Lotti, M., Organophosphate polyneuropathy in chicks, *Biochem. Pharmacol.*, 45, 131, 1993.

Perlmutter, J. S., New insights in the pathophysiology of Parkinson's disease: the challenge of positron emission tomography, *Trends Neurosci.*, 11, 203, 1988.

Perry, V. H., Brown, M. C., Lunn, E. R., Tree, P., and Gordon, S., Evidence that very slow wallerian degeneration in C57BL/Ola mice is an intrinsic property of the peripheral nerve, *European J. Neurosci.*, 2, 802, 1990.

Pope, C. N. and Padilla, S., Potentiation of organophosphorus-induced delayed neurotoxicity by phenylmethylsulfonyl fluoride, *J. Toxicol. Environ. Hlth.*, 31, 261, 1990.

Pope, C., Tanaka, D., Jr., and Padilla, S., The role of neurotoxic esterase (NTE) in the prevention and potentiation of organophosphorus induced delayed neurotoxicity (OPIDN), *Chem.-Biol. Interact.*, 87, 395, 1993.

Price, D. L. and Griffin, J. W., Neurons and ensheating cells as targets of disease processes, in *Experimental and Clinical Neurotoxicology*, Spencer, P. S. and Schaumburg, H. H., Eds., Williams & Wilkins, Baltimore, 1980, 1.

Rothstein, J. D., Martin, L. J., and Kuncl, R. W., Decreased glutamate transport by the brain and spinal cord in amyotrophic lateral sclerosis, *N. Engl. J. Med.*, 326, 1464, 1992.

Singer, T. P., Castagnoli, N., Jr., Ramsay, R. R., and Trevor, A. J., Biochemical events in the development of parkinsonism induced by 1-methyl-4-phenyl-1,2,2,6-tetrahydropyridine, *J. Neurochem.*, 49, 1, 1987.

Spencer, P. S., Nunn, P. B., Hugan, J., Ludolph, A. C., Ross, S. M., Roy, D. N., and Robertson, R. C., Guam amyotrophic lateral sclerosis — parkinsonism — dementia linked to a plant excitant neurotoxin, *Science*, 237, 517, 1987.

Travis, J., Guam: deadly disease dying out, *Science,* 261, 424, 1993.

Turski, L., Bressler, K., Retting, K. J., Löschmann, P. A., and Wachtel, H., Protection of substantia nigra from MPP⁺ neurotoxicity by N-methyl-D-aspartate autogenesis, *Nature,* 349, 414, 1991.

Verschoyle, R. D. and Aldridge, W. N., Structure-activity relationships of some pyrethroids in rats, *Arch. Toxicol.,* 45, 325, 1980.

Vijverberg, H. P. M. and van den Bercken, J., Neurotoxicological effects and mode of action of pyrethroid insecticides, *Crit. Rev. Toxicol.,* 21, 105, 1990.

Wright, C. D. P., Forshaw, P. J., and Ray, D. E., Classification of the actions of ten pyrethroid insecticides in the rat, using the trigeminal reflex and skeletal muscle as test systems, *Pest. Biochem. Physiol.,* 30, 79, 1988.

Chapter 3

TOXICITY OF OXYGEN

Donald J. Reed

CONTENTS

I. Introduction 36

II. Origin of Oxidative Stress 37

III. Cellular Injury by Toxic Oxygen Species
Generated by Redox Cycling 40

IV. Lipid Peroxidation and Antioxidant Interactions
against Oxygen Toxicity 42

V. Glutathione as a Cellular Protective Agent 49

VI. Glutathione Redox Cycle and Oxygen-induced Toxicity 51

VII. Extracellular Ca^{2+} Omission: A Model for Oxidative Stress 53

VIII. Mitochondrial GSH 55

IX. Mitochondrial Permeability Transition 57

X. Oxidation of Pyridine Nucleotides during
Permeability Transition 60

Acknowledgments 61

References 61

The abbreviations used are: αTH, α-tocopherol; αTQ, α-tocopheryl quinone; BCNU, 1,3-bis(2-chloroethyl)-1-nitrosourea; SPC, soybean phosphatidylcholine; DPPC, dipalmitoyl-phosphatidyl choline; HPLC, high performance liquid chromatography; MDA, malondialdehyde. The term "lipid oxy-radicals" refers to both lipid peroxyl radicals (LOO·) and lipid alkoxyl radicals (LO·). The former species are assumed to predominate during uninhibited lipid oxy-radical propagation. However, relative contributions of both species to αTH depletion may depend on αTH concentration (vide infra).

I. INTRODUCTION

Studies on the mechanisms of oxygen toxicity have led to a greater understanding of oxidative stress-induced toxicity. In mammalian cells, metabolism of a chemical to one or more biological reactive intermediates often can be accompanied by the loss of control of endogenous oxidative events related to the cellular requirement for molecular oxygen. Such a process, which is known as oxidative stress, may occur to an extent that ranges from a minor to a major contribution to overall toxicity of a particular chemical. For example, chemicals that are known to undergo redox cycling can cause oxidative stress to such a degree that they play a major role in chemically induced toxicity. If a chemical can form one or more adducts with cellular constituents, particularly proteins and nucleic acids, then the mechanism of toxicity may involve both the consequence of oxidative stress and adduct formation. Several mechanisms of oxygen toxicity exist in which the status of cellular integrity and protective systems are challenged by the production of reactive oxygen species.

This chapter will not describe the current work on genetic regulation by reactive oxygen species or molecular oxygen itself. Recent findings are very exciting but cannot be given adequate discussion here since many investigations are under way to describe oxygen-regulated events. For example, a recent study by Cowan and co-workers has identified oxygen responsive elements in the 5′ flanking region of the human glutathione peroxidase gene (Cowan et al., 1993).

We will discuss oxygen toxicity that is the result of oxidative stress (Sies, 1985) related to the metabolism of molecular oxygen. The continual formation of reactive oxygen species is important for normal physiological functions. However, when generated in excess they can be toxic and even more so in the presence of transition metal ions such as iron or copper (for a review see Halliwell and Gutteridge, 1990). Under normal conditions, endogenous oxidative stress does not necessarily place biological tissues and cells at risk since protective systems are present and functioning. However, protective systems can be overwhelmed by the presence of xenobiotics as well as conditions of anoxia and omission of extracellular calcium. It is the excess generation of reactive oxygen species within tissues that can cause damage to cellular constituents. In these instances, oxygen metabolism is involved with the formation of oxygen-containing reactive intermediates, many of which can undergo a cycling process between oxidation and reduction. This process is known as redox cycling and causes oxidative stress in cells. Such oxidative stress can cause oxidation of cellular constituents such as low molecular weight thiol-containing compounds including glutathione, protein thiols, as well as other functional groups on macromolecules, and peroxidation of lipids and other susceptible cellular constituents. It is now known that many chemicals can undergo both bioactivation to form biological reactive intermediates that bind to macromolecules and also enhance the formation of oxygen radicals with a concomitant oxidative stress. The evolution of bioactivation processes that form biological reactive intermediates is thought to have necessitated the concomitant evolution of cellular protection systems for cell survival in an oxygen-containing atmosphere. All tissues and cells contain systems for detoxification of biological reactive intermediates and to prevent or limit cellular damage due to oxidative stress events. In many of these cases, disruption of intracellular Ca homeostasis is implicated in the injury and conditions have been identified in which disrupted Ca^{2+} homeostasis is involved with oxidative stress. Whether oxidative stress results in loss of Ca^{2+} homeostasis or disrupted Ca^{2+} homeostasis causes oxidative stress is not known. However, the two processes appear related in some way.

Toxic processes have reversible and irreversible features that are a consequence of the interplay of bioactivation and oxidative stress with the cellular protective systems that prevent or limit cellular toxicity and possibly repair. Reversible toxicity occurs even with chemicals known as "safe" chemicals. Dose is the determining factor of whether any chemical causes irreversible toxicity. Irreversible toxicity may cause cell death regardless of what antidotal or preventive measures are taken after exposure of cell or tissues to a sufficient dose of the toxic chemical. Thus, the actual time of death of a cell is difficult to assess, but, in general terms, death can be described as being at the time that loss of cellular integrity occurs to such a degree that free exchange between the intracellular constituents and the surrounding milieu prevents cell survival. We review here oxidative stress and the mechanisms by which cells are protected by preventing or limiting cellular damage.

II. ORIGIN OF OXIDATIVE STRESS

Major contributors to oxygen toxicity are the sources of increased oxidative stress that include increased endogenous superoxide generation, loss of integrity of heme proteins or metalloproteins to enhance the contribution of transition metals to superoxide production, and inadequate antioxidant defenses based on glutathione and vitamins C and E (Table 3.1). The interaction of vitamin E, ascorbic acid, and glutathione against oxidative damage has been reviewed (Reed, 1992).

TABLE 3.1
Factors Responsible
for Initiating
Oxidative Stress

Superoxide and hydrogen peroxide
Redox cycling agents
Transition metals
Excess consumption of antioxidants

TABLE 3.2
Specific Sources of Reactive Oxygen Species Involved in Pathogenic Oxidative Stress

Endogenous

Mitochondrial electron transport system
Stimulated phagocytic cells
Non-phagocytic cells being promoted or during metabolism of xenobiotics
Induction of pro-oxidant enzymes including xanthine oxidase
Quinone semiquinone redox cycling agents
Induction of fatty acid CoA oxidases by peroxisome proliferators
Ischemia/reperfusion
Inhibition of antioxidant enzymes

Exogenous

Pollutants
Pesticides
Drugs
Ionizing radiation

Specific sources of reactive oxygen species involved in pathogenic oxidative stress are shown in Table 3.2. Without mitochondrial antioxidant defense systems such as the glutathione redox cycle, aerobic metabolism, an essential process for life in many species, would likely be impossible due to the endogenous formation of reactive oxygen species (ROS). It is estimated that nearly 90% of the total O_2 consumed by mammalian species is delivered to mitochondria where a 4-electron reduction to H_2O by the respiratory chain is coupled to ATP synthesis (Chance et al., 1979; Cadenas, 1989). Nearly 4% of mitochondrial O_2 is incompletely reduced, due to leakage of electrons along the respiratory chain, forming ROS such as the superoxide radical (O_2.), hydrogen peroxide (H_2O_2), singlet oxygen, and hydroxyl radical (HO˙) (Chance et al., 1979; Cadenas, 1989).

The superoxide radical undergoes disproportionation to H_2O_2 and O_2 enzymatically via the Mn-containing superoxide dismutase as well as nonenzymatically, although at a rate approximately four orders of magnitude less at pH 7.4 (Fridovich, 1974). Richter (1988) calculated that during normal metabolism, one rat liver mitochondrion produces 3×10^7 superoxide radicals per day. It is estimated that superoxide and hydrogen peroxide steady state concentrations are in the picomolar and nanomolar range, respectively (Forman and Boveris, 1982). Jones et al. (1981) has estimated the hepatocyte steady state H_2O_2 concentration to be up to 25 μM. If endogenously formed H_2O_2 is not detoxified to H_2O, formation of the HO˙ by a metal (iron)-catalyzed Haber Weiss or Fenton reaction can occur (Equation 1a,b) (Pryor, 1986). The HO˙ species is one of the most reactive and short-lived biological radicals and has the potential to initiate lipid peroxidation of biological membranes (Pryor, 1986; Bindoli, 1988) although not as effectively as other radicals including the ROO˙ radical (Dix and Aikens, 1993). Unless termination reactions occur, the process of lipid peroxidation will propagate, resulting in potentially high levels of oxidative stress. Therefore, detoxification of endogenously produced H_2O_2 is critical for redox maintenance of mitochondrial as well as cellular homeostasis.

A. Haber Weiss Reaction

$$Fe^{+3} + O_2^- + H_2O_2 \rightarrow Fe^{2+} + O_2 + OH^- + HO˙ \tag{1a}$$

B. Fenton Reaction

$$Fe^{+2} + H_2O_2 \rightarrow Fe^{+3} + HO˙ + OH^- \tag{1b}$$

Goldstein et al. (1993) have reviewed the Fenton reagents and concluded that the oxidizing intermediates formed via the Fenton-like reactions *in vivo* are not possible to identify. Also, the question of whether an ˙OH radical is formed rather than a metal in a higher oxidation state that is capable of causing biological damage is nearly impossible to determine.

Sohal (1993) has concluded that there is a variation in the sites of superoxide or hydrogen peroxide generation among mitochondria from different tissues and species. This is a result of the rate of mitochondrial superoxide production being dependent on at least three variables: (a) ambient oxygen concentration; (b) levels of autoxidizable respiratory carriers, especially ubiquinone; and (c) the redox state of the autoxidizable carriers (Sohal, 1993; Turrens and McCord, 1990; Boveris and Cadenas, 1982; Forman and Boveris, 1982; Chance et al., 1979).

Paraidathathu et al. (1992) have shown that mitochondria from hypoxic rat heart tissue produced less reactive oxygen species than normoxic tissue. However, they found that in the presence of calcium the reactive oxygen species of mitochondria from hypoxic rats increased to that of normoxic rats. These workers concluded that hypoxic injury may be a result of the combination of production of reactive oxygen species and increased intracellular calcium levels by mechanisms that may not be mutually exclusive.

Since mitochondria contain the enzymes and cofactors necessary for the GSH/GSSG redox cycle (Flohe and Schlegel, 1971) but do not contain catalase (Neubert et al., 1962), we may assume that a primary function of mitochondrial glutathione (GSH) is for the detoxification of endogenously produced H_2O_2. This redox cycle requires the enzymes GSH peroxidase (selenium-containing) and GSSG reductase along with the cofactors GSH and NADPH (Figure 3.1). In addition to detoxifying H_2O_2, GSH also protects protein sulfhydryls from oxidation (Vignais and Vignais, 1973).

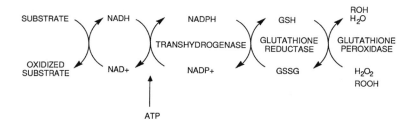

FIGURE 3.1.   Mitochondrial glutathione and the glutathione redox cycle.

Stimulated phagocytic cells including polymorphonuclear leukocytes (PMNs) and macrophages can generate large quantities of superoxide as a mechanism for destruction of foreign cells. Such an oxidative burst requires that these cells maintain defenses against their own oxygen toxicity. It is known that an NADPH oxidase provides the catalysis for the rapid consumption of molecular oxygen (Babior, 1984, 1987).

Non-phagocytic cells can be promoted by agents such as TPA that can induce oxidative events that may even involve the formation of xanthine oxidase by the conversion of the dehydrogenase to an oxidase (Frenkel, 1992). Exogenous redox cycling agents include the drug, Adriamycin® and the herbicide, paraquat. Such agents have the potential of converting large quantities of oxygen to superoxide by undergoing one-electron redox cycling — the reduced form of the redox agent being able to provide a one-electron reduction of oxygen.

III. CELLULAR INJURY BY TOXIC OXYGEN SPECIES GENERATED BY REDOX CYCLING

Diquat (DQ), a bipyridyl herbicide that is hepatotoxic, is a model compound for studies of redox cycling. Diquat utilizes molecular oxygen to generate large amounts of superoxide anion radical and hydrogen peroxide within cells (Figure 3.2). Evidence to support the protective role of the glutathione redox cycle is that the toxicity of diquat *in vivo* and *in vitro* appears to be mediated by redox cycling and is greatly enhanced by prior treatment with BCNU (Sandy et al., 1987), which is an inhibitor of glutathione reductase (Reed, 1985). Ebselen, a synthetic compound possessing glutathione peroxidase-like activity, protects against diquat cytotoxicity when extracellular glutathione is present in the medium (Muller et al., 1984). Superoxide anion radical, which can reduce ferric iron to ferrous iron, is produced during diquat toxicity (Keberle, 1964). Desferrioxamine, which chelates intracellular iron in the ferric state with an affinity constant of 10^{31}, provides considerable protection against diquat-induced toxicity. Therefore, hydrogen peroxide and transition metals have been suggested as major contributors to diquat toxicity. Even though the hydroxyl

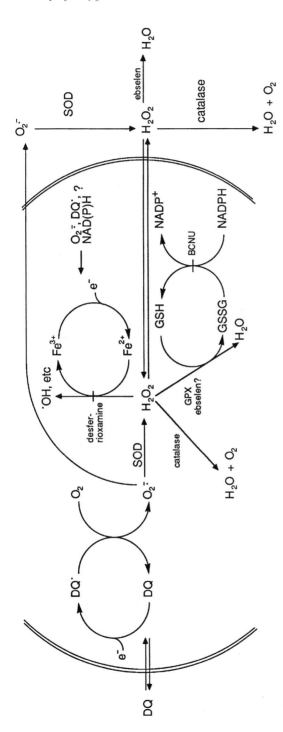

FIGURE 3. 2. Scheme of postulated interactions between diquat, diquat-derived active oxygen species, internal and external enzymes, and transition metals. DQ, diquat; DQ·, diquat radical; GPX, glutathione peroxidase.

radical or a related species seems the most likely ultimate toxic product of the hydrogen peroxide/ferrous iron interaction, scavengers of hydroxyl radical afford only minimal protection. The high degree of reactivity of hydroxyl radicals assures, however, that the site of interaction with cellular components is within close proximity (a few angstroms) of the site of generation of the radical. Much remains to be understood about the mechanism of cytotoxicity by the various redox cycling agents including the quinones, menadione, and Adriamycin.

IV. LIPID PEROXIDATION AND ANTIOXIDANT INTERACTIONS AGAINST OXYGEN TOXICITY

Interaction of antioxidants provides essential protection against the damaging effects of an oxygen-based metabolism (Table 3.3) (for a review see Di Mascio et al., 1991). Complementary antioxidant systems in cytoplasmic and membrane compartments provide efficient cellular defense against oxidative injury (Sies, 1985). αTH[1] protects cellular membranes against oxidative damage (Machlin, 1980; Burton et al. 1983; Tappel, 1972), and dietary αTH deficiency enhances the susceptibility of biological membranes to oxidative damage *in vitro* (Tappel, 1972; McCay et al., 1971; Kornbrust and Mavis, 1980; Sevanian et al., 1982) and *in vivo* (Litov et al., 1981; Herschberger and Tappel, 1982; Dillard et al., 1982). Nonetheless, clinical manifestations of αTH deficiency in humans are poorly defined and are usually secondary to other disease states (Machlin, 1980; Muller et al., 1983). Diplock (1991) and Rice-Evans and Diplock (1993) have reviewed the status of antioxidant nutrients and disease prevention. This section will focus on the interaction of antioxidants for protection against oxidative damage.

Lipid peroxidation, which can cause extensive damage to subcellular organelles and biomembranes, has been demonstrated to occur in isolated mitochondria, lysosomes, microsomes (Tappel, 1973), and nuclei (Tirmenstein and Reed, 1988). Lipid peroxidation is an important biological consequence of oxidative cellular damage (Plaa and Witschi, 1976) that can be measured by several noninvasive techniques in humans (for a review see Pryor and Godber, 1991). The destruction of unsaturated fatty acids, which occurs in lipid peroxidation, has been linked with altered membrane structure (Eichenberger et al., 1982) and enzyme inactivation (Baker and Hemsworth, 1978; Thomas and Reed, 1990). In addition to lipid hydroperoxides and lipid radicals (Porter, 1984), lipid peroxidation generates activated oxygen species such as hydroxyl

TABLE 3.3
Protection Against Oxygen Toxicity

Superoxide scavengers
Transition metal chelation
Enhancement of antioxidant levels
Enhancement of energy sources for antioxidant maintenance

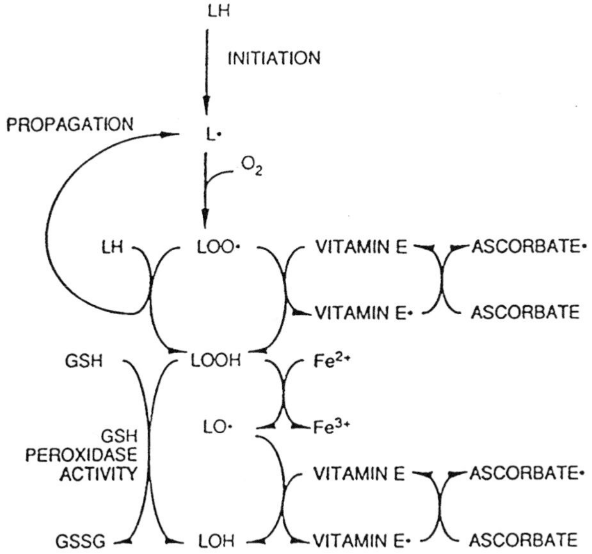

FIGURE 3.3. Protective interactions of vitamin E, ascorbate, and glutathione against lipid peroxidation.

radicals (O'Brien and Hawco, 1978) and superoxide anions (Svingen et al., 1978). The decomposition of peroxidized polyunsaturated fatty acids also generates reactive carbonyl compounds such as malondialdehyde (MDA) and hydroxyalkenals (Brambilla et al., 1986).

Biochemical regulation of cellular αTH status is only beginning to be understood. Several studies have addressed the hypothesis that soluble cellular antioxidants, particularly ascorbic acid, maintain membrane αTH levels by regenerating αTH from its oxidation products (Figure 3.3). Indeed, the reduction of the αTH semiquinone radical by ascorbic acid was reported to proceed in homogenous solution with a bimolecular rate constant of 1.55×10^6 M^{-1} s^{-1} (Packer et al., 1979). Synergistic antioxidant protection by αTH and ascorbic acid has been demonstrated in lard (Golumbic and Mattill, 1941), homogenous solution (Niki et al., 1983), micellar (Tappel et al., 1961; Barclay et al., 1985), and liposome suspensions (Leung et al., 1981; Fukuzawa et al., 1985; Doba et al., 1985; Niki et al., 1985). Other reports have provided direct evidence for the αTH-sparing action of ascorbic acid (Bascetta et al., 1983; Scarpa et al., 1984). Although these data are consonant with a hypothesis of reductive regeneration of αTH, the reaction of ascorbic acid as well as αTH with free radicals complicates interpretation and may account for the additive or synergistic antioxidant effects observed (McCay, 1985). In biological lipid peroxidation, soluble antioxidants also may exert pro-oxidant effects via reduction of metals (Aust et al., 1985). The role of soluble antioxidants thus appears complicated. αTH is principally consumed by lipid oxy-radicals rather than by oxygen radical initiators. Lipid oxy-radical propagation is effectively inhibited by αTH at concentrations above a threshold level. The celebrated Fenton reaction (Equations 1b–4) (Aust et al., 1985) can participate in lipid peroxidation events as shown below:

$$Fe^2 + H_2O_2 \rightarrow Fe^{3+} + \cdot OH + OH^- \tag{1b}$$

$$LH + \cdot OH \rightarrow L\cdot + H_2O \tag{2}$$

$$L\cdot + O2 \rightarrow LOO\cdot \tag{3}$$

$$LOO\cdot + LH \rightarrow LOOH + L\cdot \tag{4}$$

A Fe^{2+} and lipid hydroperoxide-dependent pathway also occurs (Aust et al., 1985; Svingen et al., 1979) (Equations 5 and 6).

$$Fe^{2+} + LOOH \rightarrow Fe^{3+} + LO\cdot + OH^- \tag{5}$$

$$LO\cdot + LH \rightarrow LOH + L\cdot \tag{6}$$

Both ascorbic acid and GSH act as pro-oxidants when incubated with αTH-free liposomes and Fe^{2+}/H_2O_2. The pro-oxidant effect of ascorbic acid, but not of GSH, is reversed by incorporation of αTH into the liposomes. αTH levels in excess of the threshold are effectively maintained by ascorbic acid, but GSH antagonizes the antioxidant effects of the αTH/ascorbic acid combination. Effective antioxidant protection by αTH appeared to be due to efficient reaction with lipid oxy-radical in the bilayer rather than to interception of initiating oxygen radicals. At concentrations above a threshold level of approximately 0.2 mole percent (mol αTH per mol phospholipid × 100), αTH completely suppressed lipid oxy-radical propagation, which was measured as malondialdehyde production. Both ascorbic acid and glutathione, alone or in combination, can enhance lipid oxy-radical propagation (Liebler et al., 1986). αTH, incorporated into membranes such as in liposomes at concentrations above its threshold protective level, can reverse the pro-oxidant effects of 0.1–1.0 mM ascorbic acid but not those of glutathione. Ascorbic acid also prevented αTH depletion. The combination of ascorbic acid and subthreshold levels of αTH only temporarily suppresses lipid oxy-radical propagation and does not maintain the αTH level. Glutathione antagonizes the antioxidant action of the αTH/ascorbic acid combination regardless of αTH concentration. These observations indicate that membrane αTH status can control the balance between pro- and antioxidant effects of ascorbic acid. These studies also provide direct evidence that ascorbic acid interacts directly with components of the phospholipid bilayer (Liebler et al., 1986).

The remarkable antioxidant efficiency of the αTH/ascorbic acid combination may be due in part to efficient radical scavenging by these agents in hydrophobic and hydrophilic environments, respectively. Liebler et al. (1986) have shown that ascorbic acid also regenerates αTH from its oxidation products. In addition, αTH completely reversed the pro-oxidant effects of 0.1 mM or 1.0 mM ascorbic acid. The inhibition of lipid oxy-radical propagation coincided with maintenance of αTH. The ability of ascorbic acid, a relatively polar and lipid insoluble molecule, to interact directly with components of the lipid bilayer has been questioned (McCay, 1985). However, the data of Scarpa et al. (1984) indicates direct reaction between ascorbic acid and liposomal oxy-radicals.

The limited antioxidant capacity of the αTH/ascorbic acid combination at subthreshold αTH concentrations is noteworthy. αTH at 0.1 mole percent (mol αTH per mol phospholipid × 100) does not suppress lipid oxy-radical propagation in liposomes unless 0.5–1.0 mM ascorbic acid is present. Even so, αTH status was not maintained — the rate of αTH consumption was merely decreased. The threshold level thus represents a point of no return — propagation reactions cannot be effectively suppressed, nor can the αTH level be maintained. Ascorbic acid, even at 1 mM, cannot maintain αTH levels because the rate of reaction of the αTH semiquinone radical with lipid oxy-radicals exceeds the rate of its reduction to αTH.

Oxidative challenge in liposomes may mimic biological oxidative injury, in which oxygen radicals are generated due to the presence of copper and iron (Aust et al., 1985; Miller et al., 1990). In the absence of αTH, the availability of Fe^{2+} determines the extent of SPC liposome peroxidation. Reduction of Fe^{3+} by ascorbic acid or GSH maintains an active Fe^{2+} pool and enhances both the rate and extent of lipid oxy-radical propagation. The concentration-dependent pro-oxidant effect of GSH is thus easily explained as described by Miller et al. (1990). The pro-oxidant action of ascorbic acid, however, was maximal at the lowest concentration studied (0.1 mM) and concentrations above 1 mM temporarily inhibited lipid peroxidation (Liebler et al., 1986). Inhibition due to scavenging of hydroxyl radicals — presumably the initiating species in this model — by ascorbic acid is unlikely. GSH produces no such inhibition despite being a superior hydrogen atom donor (Willson, 1983). On the other hand, ascorbic acid, at higher concentrations, may react directly with lipid oxy-radicals at the lipid-water interface. Because GSH displays no antioxidant activity with liposomes, it does not appear to react with lipid oxy-radicals to a significant extent or regenerate αTH from its oxidation products. Both reactions may be kinetically unfavorable for GSH (Willson, 1983). The notable antagonism of the αTH/ascorbic acid antioxidant combination by GSH may involve at least two mechanisms. First, GSH chelation and reduction of iron enhances and sustains the oxidative challenge (Bucher et al., 1983). Second, the glutathione radicals thus produced are themselves rapidly reduced by

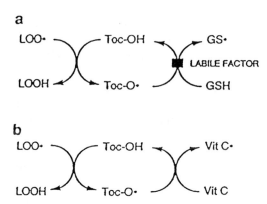

FIGURE 3.4. Antioxidant interactions: (a) regeneration of vitamin E (Toc-OH) via GSH-dependent labile factor; (b) ascorbic acid (vitamin C) regeneration of vitamin E. LOO·, lipid peroxyl radical; LOOH, lipid hydroperoxide.

ascorbic acid (Willson, 1983). The consumption of ascorbic acid would consequently exceed that expected on the basis of its antioxidant functions alone.

Ascorbic acid inhibits lipid peroxidation by regenerating αTH rather than by reacting directly with lipid oxy-radicals (Figure 3.4). These studies represent one of the first demonstrations of a "sparing" effect of ascorbic acid upon αTH in lipid bilayers under conditions where significant reaction between ascorbic acid and radicals can be unequivocally ruled out. Furthermore, ascorbic acid (at 0.1 mM) offers no antioxidant protection when αTH is present, but at a subthreshold level. This observation supports a regeneration hypothesis but contradicts the view that ascorbic acid reacts independently with lipid oxy-radicals to spare αTH. Liebler et al. (1989) have provided evidence that, *in vitro* or in liposomes, ascorbate does not reduce α-tocopherone directly but reacts instead with the tocopherone cation. Slow rates of these reactions have led these workers to suggest that biochemical catalysis would be required to complete a two-electron TH redox cycle in biological membranes.

A central conclusion is that the balance between pro- and antioxidant effects of ascorbic acid is critically dependent upon the bilayer concentration of αTH relative to the threshold level. Ascorbic acid cannot maintain αTH at subthreshold levels, presumably because the rate of αTH consumption exceeds the rate of its reductive regeneration from a radical intermediate by ascorbic acid.

The effectiveness of the αTH/ascorbic acid combination is thus intimately related to the threshold effect. The liposome studies indicate that the balance between pro- and antioxidant effects of ascorbic acid ultimately depends on the αTH status of the lipid bilayer. The results further imply that maintenance of membrane αTH status may be an essential function of cellular ascorbic acid *in vivo*. If so, then NADH-dependent reduction of semidehydroascorbate (Ito et al., 1981) may subject αTH recycling to a limited degree of metabolic control. Attempts to manipulate tissue αTH levels by modifying dietary ascorbic acid intake have met with limited success (McCay, 1985). However, ascorbic acid treatment enhanced lipid peroxidation in αTH-deficient rats (Litov et al., 1981), and Litov and co-workers concluded that, in the absence of αTH, lipid oxy-radical propagation in cellular membranes proceeds unchecked by ascorbic acid. This conclusion is essentially identical to that of the liposome studies; ascorbic acid, at concentrations below 1 mM, did not function as an antioxidant unless the αTH concentration exceeded the threshold level.

Damage to membrane lipids and the associated alterations in bulk properties of membranes is the basis for chemical-induced loss of cell viability. Horrobin (1991) has focused on the possibility that loss of membrane essential fatty acids is the basis for cell injury rather than the accumulation of toxic lipid peroxidation products. Possibly in agreement with that hypothesis, lipid peroxidation and further oxidative modification of low density lipoprotein (LDL) is thought to increase its atherogenicity (Retsky et al., 1993; Steinberg et al., 1989; Schwartz et al., 1991). These workers have shown that ascorbic acid oxidation products including dehydro-L-ascorbic acid protect human LDL against atherogenic modification and that anti-, rather than pro-oxidant activity, was observed for ascorbic acid in the presence of transition metal ions (Retsky et al., 1993).

The failure of GSH to function as an antioxidant in liposomes despite mounting evidence for such a role in biological systems (Haenen and Bast, 1983; McCay et al., 1989) is significant. Enzymic participation in the antioxidant action of GSH is thought to be essential, and the results of these studies reinforce that view (Figure 3.4). Evidence for protein- and GSH-dependent antioxidant protection of cell membranes has been reported (Reddy et al., 1982; Hill and Burk, 1984; Haenen and Bast, 1983), but evidence for the participation of αTH in GSH-dependent protection is somewhat equivocal. Further study of GSH-dependent antioxidant activity in biological membranes may reveal a system whose function complements that of αTH.

One possible mechanism by which GSH would stimulate lipid peroxidation is by reducing iron and thus promoting Fe^{2+}-dependent peroxidation. This is not a new suggestion; it is well supported by previously published work (see Misra, 1974; Rowley and Halliwell, 1982). Moreover, evidence that thiols may directly reduce iron and stimulate peroxidation has come from Aust and colleagues (Bucher et al., 1983; Tien et al., 1982).

Vitamin E participates in the inhibition of lipid peroxidation of membranes by breaking of the chain propagation (Niki et al., 1991) by reactions that appear to include scavenging of lipid hydroperoxyl radicals. Murphy et al. (1989) have shown that the protection by ascorbate and glutathione against microsomal lipid peroxidation is dependent on vitamin E. Loss of vitamin E is related to both chemiluminescence and thiobarbituric-reactive material formation which increased when the αTH content of the microsomes had decreased to about 0.38 nmol/mg protein, which is about 86% of the initial content (Murphy et al., 1989).

Dietary supplementation with vitamin E gave a 10- to 20-fold increase of vitamin E in the rat liver mitochondrial membrane (Maguire et al., 1989).

Treatment of submitochondrial particles with an oxidizing system composed of lipoxygenase and arachidonic acid provided evidence that reduction of the tocopheroxyl radical could occur by NADH, succinate, and reduced cytochrome c-linked oxidation. Thus, the electron transport system is proposed to have an important physiological role in recycling vitamin E by reduction of the tocopheroxyl radical to prevent its accumulation and vitamin E consumption.

A limited number of studies have focused on the susceptibility of the cell nucleus to lipid peroxidation. The nuclear membrane regulates the transport of mRNA into the cytoplasm and aids in the process of nuclear division. DNA is also frequently associated with certain regions of the nuclear membrane (Franke, 1974), and it seems likely that nuclear membrane peroxidation may disrupt many of these critical functions. The proximity of the nuclear membrane to DNA could also contribute to the interaction of DNA with reactive compounds generated in lipid peroxidation. Several studies indicate that such lipid peroxidation products can alter the structure and function of DNA (Akasaka, 1986; Ueda et al., 1985; Brawn and Fridovich, 1981; Mukai and Goldstein, 1976). This fact is of importance since hydroxyl radicals diffuse an average of only 60 angstroms before reacting with cellular components (Roots and Okada, 1975). Assays for 8-hydroxy-2'-deoxyguanosine as a biomarker of oxidative DNA damage includes *in vivo* studies with urine samples (Shigenaga and Ames, 1991). Nuclear peroxidation may also increase interactions between more stable peroxidation products and DNA. The cytosolic enzymes aldehyde dehydrogenase (Hjelle and Petersen, 1983), glutathione transferase (Alin et al., 1985), and glutathione peroxidase (Christophersen, 1968) have all been shown to metabolize various reactive lipid peroxidation products. Such cytosolic enzymes may metabolize peroxidation products generated throughout the cell before they diffuse into the nucleus and interact with DNA.

Endogenous αTH levels in isolated rat liver nuclei have been measured and found to be 0.045 mol E (mol αTH per mol phospholipid $\times$ 100) (Tirmenstein and Reed, 1989). This value corresponds to 970 polyunsaturated fatty acid (PUFA) moieties to one molecule of αTH in the nuclear membrane. These values are higher than values reported for rat liver microsomes (3313) (Sevanian et al., 1982) and mitochondria (2100) (Gruger and Tappel, 1971). A threshold level of 0.085 mol % for the prevention of NADPH-induced lipid peroxidation was established for isolated nuclei. That value could be lowered to 0.040 mol % when 1 mM GSH was added to assist in the inhibition of lipid peroxidation. The ability of GSH to enhance αTH-dependent protection against nuclear lipid peroxidation appears to be mainly by a "sparing" effect on the near threshold level of αTH in the nuclear membrane.

Thiol groups are well known to be important for normal protein functions and increasing evidence supports the vital importance of these thiols for cell viability during cytotoxic events. Membrane-bound enzymes are damaged during lipid peroxidation, and evidence of vitamin E protection strongly supports a free radical mechanism for protein damage via oxidative stress (Dean and Cheeseman, 1987; Thomas and Reed, 1990). Oxidative stress can cause loss of protein functions by damaging amino acid residues other than cysteine including methionine, tryptophan, and histidine. An important aspect of such damage is that lipid peroxidative events can amplify free radical processes which propagate chain reactions. Failure to terminate free radical processes with a chain-breaking antioxidant, such as vitamin E, can lead to 4 to 10 propagation events occurring per initiation and thus each initiation is amplified (McCay et al., 1976). Products of lipid peroxidation are toxic and lipid peroxidation events can also compromise detoxication systems. Since the reduction of lipid hydroperoxides by GSH utilizes NADPH for the regeneration of GSH from GSSG, the rate of NADPH production can be limiting during oxidative stress. Therefore GSSG may be transported from the liver when not reduced due to limited levels of NADPH. Decreased availability of NADPH and GSH can impair other GSH-dependent detoxication pathways including metabolism of hydrogen peroxide, decreased protection of thiols in protein (Pascoe and Reed, 1987), and decreased reaction with free radicals. Thus, energy-dependent processes involving NADPH, GSH, and thiols in proteins, appear to be critically involved in cellular homeostasis during drug-induced toxicity.

The prevention of nitrofurantoin-induced cytotoxicity in isolated hepatocytes by fructose appears to relate to maintenance of the cellular glutathione pool by limiting GSSG efflux from these cells. Silva et al. (1991) have concluded that fructose induced ATP depletion (20% of control) to such an extent that ATP-dependent GSSG efflux was prevented, thus, protecting the cells. This protection is by the retained GSSG being reduced back to GSH for limiting the nitrofurantoin-induced oxidative stress.

In addition, a combination of calcium release from intracellular stores and inhibition of calcium efflux appears to result in a marked and sustained increase in cytosolic concentration of calcium associated with surface blebbing and increased activity of certain calcium-dependent enzymes.

V. GLUTATHIONE AS A CELLULAR PROTECTIVE AGENT

Mammalian cells have evolved a major protective system to minimize injurious events that result from biotransformation of xenobiotics and normal oxidative products of cellular metabolism. This system is dependent upon the unique tripeptide, glutathione (GSH). Depending on the cell type, the intracellular concentration of GSH is in the range of 0.5 to 10 mM (for a review see Reed, 1990). Concentrations in mammalian liver are 4 to 8 mM with nearly all of the glutathione being present as reduced glutathione, GSH, and less than 5% of the total glutathione being present as oxidized glutathione, glutathione disulfide, GSSG, with minor fractions being mixed disulfides of GSH and other cellular thiols, and minor amounts of thioethers from endogenous conjugation reactions (Kosower and Kosower, 1978). Overall, the GSH content of various organs and tissues represents at least 90% of the total nonprotein, low molecular weight thiols. GSH can be depleted directly by conjugation with electrophiles and indirectly by the addition of inhibitors of GSH biosynthesis and regeneration (Figure 3.5). The liver appears unique in that GSH is synthesized *in vivo* with the sulfur of cysteine supplied by methionine via the cystathionine pathway (Reed and Beatty, 1980). Cysteine formed by this pathway as well as the low level of cysteine that is transported into liver cells serve as substrate for γ-glutamylcysteine synthase to form γ-glutamycysteine which is then coupled with glycine via GSH synthetase (EC 6.3.2.3) to complete the synthesis of GSH. Most of the other cell types with few exceptions must depend entirely on the uptake of cysteine because of the lack of the cystathionine pathway for cysteine synthesis and in turn GSH synthesis. Inhibition of γ-glutamylcysteine synthetase (EC 6.3.2.2), the cytosolic rate limiting enzyme, by buthionine sulfoximine (BSO) diminishes the rate of GSH synthesis but does not cause depletion of mitochondrial GSH during cytosolic GSH depletion. The two main functions of GSH are as a nucleophilic "scavenger" of numerous compounds and their metabolites, via enzymatic and chemical mechanisms, converting electrophilic centers to thioether bonds, and as a substrate in the GSH peroxidase-mediated metabolism of hydroperoxides to

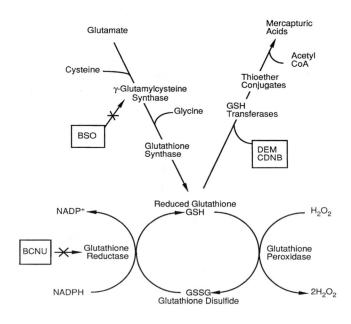

Adapted from Harlan et al., 1984

FIGURE 3.5. Biosynthesis and depletion of GSH and the glutathione redox cycle.

the corresponding alcohols. Alternatively, GSH content in isolated hepatocytes can be augmented by the addition of its sulfur-containing precursors, such as cysteine (or N-acetyl cysteine) or methionine, to maximize biosynthesis. Enhanced GSH synthesis has major beneficial effects during severe GSH depletion which is caused by consumption of GSH during xenobiotic metabolism. Cystine has a sparing effect on the requirement of the essential amino acid methionine in the rat (Womack et al., 1937). This observation is in agreement with the unidirectional process of trans-sulfuration in which methionine sulfur and serine carbon are utilized in cysteine biosynthesis via the cystathionine pathway.

The cystathionine pathway is of major importance to pathways of drug metabolism in the liver that involve GSH or cysteine, or both. Depletion of GSH by rapid conjugation can increase synthesis of GSH to rates as high as 2 to 3 μmol/hr per g wet liver tissue (White, 1976). The cysteine pool in the liver, which is about 0.2 μmol/g, has an estimated half-life of 2 to 3 min at such high rates of synthesis of GSH. Although the cystathionine pathway appears to be highly responsive to the need for cysteine biosynthesis in the liver, the organ distribution of the pathway may be limited. In mammals, such as rats, the liver may be the main site of cysteine biosynthesis, which occurs via the cystathionine pathway. Maintenance of high concentrations of GSH in the liver, in association with high rates of secretion into plasma and extensive extracellular degradation of GSH and GSSG, supports the concept that liver GSH is a physiological reservoir of cysteine (Tateishi et al., 1977; Higashi et al., 1977).

In vivo treatment of rats with AT-125, an inhibitor of γ-glutamyl transpeptidase, prevents degradation of GSH in plasma leading to massive urinary excretion of GSH (Reed and Ellis, 1982). This treatment also lowers the hepatic content of GSH because it inhibits recycling of cysteine to the liver. A physiologic decrease in interorgan recycling of cysteine to the liver for synthesis of GSH also may account in part for the decrease of hepatic GSH during starvation and for the marked diurnal variation in concentration of GSH in liver. The efflux of liver GSH and metabolism of the resulting plasma GSH and GSSG appears to help ensure a continuous supply of plasma cysteine. This cysteine pool should in turn minimize the degree of fluctuation of GSH concentrations within the various body organs and cell types that require only cysteine or cystine, or both, rather than methionine for synthesis of GSH.

VI. GLUTATHIONE REDOX CYCLE AND OXYGEN-INDUCED TOXICITY

The intracellular concentration of GSH in isolated hepatocytes has been examined under conditions that result in oxidative stress. Production of malondialdehyde, which is an index of lipid peroxidation and occurs during oxidative stress, can be stimulated by addition of a glutathione depletor, diethyl maleate. This observation suggests that intracellular concentrations of GSH under these conditions are important for membrane and cellular integrity. That is, GSH protects against oxidative damage to unsaturated fatty acids in biological membranes.

A major endogenous protective system against reactive oxygen species is the glutathione redox cycle (for a review see Reed, 1986). GSH depletion to about 20 to 30% of normal level of glutathione can impair the cell's defense against the toxic actions of both biological reactive intermediates and reactive oxygen species and may lead to cell injury and death. Endogenous oxidative stress, which is a normal physiological process, is a consequence of aerobic metabolism that occurs mostly in the mitochondria of eukaryote cells. If not decomposed, H_2O_2 can lead to the formation of the very reactive hydroxyl radical and cause the formation of lipid hydroperoxides that can damage membranes, nucleic acids, proteins, and alter their functions.

GSH is maintained in a redox couple with GSSG within the cell and is regenerated by GSH reductase (EC 1.6.4.2), a cytosolic NADPH-dependent enzyme. Inhibition of this enzyme, and hence of GSH regeneration, with 1,3-bis(2-chloroethyl)-1-nitrosourea (BCNU) also depletes intracellular GSH (Figure 3.5). It is apparent that a major protective role against the reactive drug intermediates, which are generated by bio-reduction and cause oxidative stress by redox cycling, is provided by the ubiquitous glutathione redox cycle (Figure 3.5). This cycle utilizes NADPH– and, indirectly, NADH-reducing equivalents in the mitochondrial matrix as well as the cytoplasm to provide GSH by the glutathione reductase-catalyzed reduction of GSSG. When the glutathione redox cycle is functioning at maximum capacity to eliminate hydrogen peroxide, a regulatory effect is imposed on other NADPH-dependent pathways.

Glutathione peroxidase is a selenium-dependent enzyme that is extremely specific for glutathione and is capable of rapidly detoxifying hydrogen peroxide and certain hydroperoxides. Selenium-dependent glutathione peroxidase activity is the result of the expression of multiple isozymes. Four isozymes have been characterized and they are: (a) the classical cellular glutathione peroxidase, GSHPx-1, (b) the phospholipid hydroperoxide glutathione peroxidase, GSHPX, (c) the plasma glutathione peroxidase, GSHPx-P, and (d) GSHPx-GI (Chu et al., 1993). Weitzel and Wendel (1992) have reported findings that phospholipid hydroperoxide-glutathione peroxidase activity regulates the activity of 5-lipoxygenase via regulating the tone of endogenous hydroperoxides.

Ursini et al. (1985) have purified an interfacial glutathione peroxidase. This enzyme has been shown to reduce linoleic acid hydroperoxides, cumene hydroperoxide, tert-butyl hydroperoxide, and hydrogen peroxide. However, this enzyme, which does not conjugate CDNB with GSH (Ursini et al., 1985), displays glutathione peroxidase activity toward cumene hydroperoxide, hydrogen peroxide, and lipid hydroperoxides and is distinct from the classical glutathione peroxidase (Nakamura et al., 1974). Evidence suggests that the enzyme is interfacial in character and can interact directly with liposomes to reduce phospholipid hydroperoxides (Ursini et al., 1985). The addition of this protein to microsomal incubation mixtures inhibited lipid peroxidation (Ursini et al., 1985). Substrate specificities indicate that this enzyme is distinct from the nuclear glutathione transferase (Tirmenstein and Reed, 1989).

Lipid hydroperoxides located in biological membranes increase the potential for lipid peroxidation. If this potential is to be diminished these peroxides must be removed from the membrane environment or be reduced to lipid alcohols. If glutathione peroxidase activity is associated with the phospholipid bilayer of the nuclear membrane, such an association may contribute to the ability of the peroxidase to reduce lipid hydroperoxides. Since lipid hydroperoxides can initiate lipid peroxidation, the reduction of these compounds can contribute to the inhibition of peroxidation (Figure 3.3). The association of a glutathione-dependent peroxidase with membranes may encourage the reduction of lipid hydroperoxides located within lipid bilayers.

Glutathione protection of isolated rat liver nuclei against lipid peroxidation is abolished by exposing isolated nuclei to the glutathione transferase inhibitor S-octylglutathione (Tirmenstein and Reed, 1989). S-Octylglutathione also inhibited nuclear glutathione transferase activity and glutathione peroxidase activity. A large percentage of the glutathione transferase activity associated with isolated nuclei was solubilized with 0.3% Triton X-100®. Studies suggest that this treatment removes nuclear membranes while preserving the integrity of the remaining nucleus.

Extraction of nuclei with 1 M NaCl accompanied by brief sonication failed to solubilize the peroxidase activity. Extraction with 0.3% Triton X-100, however, solubilized the GSH dependent peroxidase activity. Electron microscopy studies conducted by Dabeva et al. (1977) indicate that treatment with even higher concentrations of Triton X-100 removed the outer nuclear membrane but preserved the integrity of the remaining nucleus. Based on this information, it appears that the peroxidase activity is associated with the nuclear membrane. This activity in conjunction with GSH may contribute to the inhibition of lipid peroxidation in nuclear membranes and thereby preserve the integrity of this important membrane system. Increasing evidence suggests that this inhibition of peroxidation may in turn protect the structure and function of DNA.

VII. EXTRACELLULAR CA^{2+} OMISSION: A MODEL FOR OXIDATIVE STRESS

An interrelationship exists in isolated rat hepatocytes between Ca^{2+} status and oxidative stress. Incubation in the absence of extracellular Ca^{2+} increases levels of thiobarbituric acid-reactive substances and has a marked loss of mitochondrial and cytosolic GSH occurring concomitant with GSSG formation and GSH efflux (Fariss et al., 1984). These results, particularly the enhanced malondialdehyde (MDA) production in hepatocytes subjected to Ca^{2+} deprivation, are highly suggestive of an oxidative stress caused by omission of extracellular Ca^{2+}. Under 95% O_2/5% CO_2, the apparent oxidative stress generated by hepatocyte incubation in Ca^{2+}-free medium leads to significant cell injury as determined by K^+ and LDH leakage. Hepatocyte cytosolic and mitochondrial GSH content, as well as vitamin E, are rapidly depleted during Ca^{2+} omission (Thomas and Reed, 1988a,b). A marked loss of protein thiols is also observed. Addition of antioxidants such as vitamin E and N,N'-diphenyl-p-phenylenediamine and the iron chelators, desferrioxamine and EDTA, prevents GSH, vitamin E, and protein thiol loss and, accordingly, prevents MDA production (Thomas and Reed, 1988a,b). Protection against cell injury (K^+ and LDH leakage) is also provided by these agents indicating that the oxidative stress is capable of inducing significant cell damage.

EGTA, a metal ion chelator which is less specific for iron and more specific for Ca^{2+}, also prevents oxidative damage and cell injury. Thus, even with extracellular Ca^{2+} lacking, Ca^{2+} can play a pivotal role in cell injury. As mitochondria are sensitive to perturbations in cellular Ca^{2+} status, a question was whether the absence of extracellular Ca^{2+} affects the mitochondria. Ca^{2+} omission results in a marked decline of $\Delta\Psi$ which can be prevented by ruthenium red and La^{3+}, which block Ca^{2+} uptake through the uniport (Thomas and Reed, 1988a,b). Ruthenium red and La^{3+} eliminate the oxidative stress and cell damage associated with the absence of extracellular Ca^{2+}. From these results it can be concluded that a disturbance in mitochondrial Ca^{2+} handling is associated with the generation of the oxidative stress induced by Ca^{2+} deprivation. An initial efflux of Ca^{2+} from hepatocytes occurs which decreases the total cell Ca^{2+} content to 20% of control. It remains to be determined, however, what dictates the efflux of mitochondrial Ca^{2+}. The ability of antioxidants to prevent the collapse of $\Delta\Psi$ suggests that oxidative stress is critical to the process.

Studies with the protonophore, carbonylcyanide-m-chlorophenylhydrazone, indicate that a pronounced loss of $\Delta\Psi$ does not necessarily lead to cell injury as assessed by LDH leakage. Furthermore, oxidative stress generated by paraquat and BCNU treatment has little effect on $\Delta\Psi$ but led to severe cell damage that is not preventable by ruthenium red (Thomas and Reed, 1988a,b). These comparative studies tend to indicate that the oxidative stress induced by an absence of extracellular Ca^{2+} is distinct from that generated by redox active chemicals in that it appears to induce Ca^{2+} cycling in/out of the mitochondria which serves to promote the oxidative stress. Cell death does not appear to result solely from a decline of $\Delta\Psi$, which may predispose these cells to oxidative damage by impairment of endogenous antioxidant defense mechanisms. Such impairment appears related to loss of reducing equivalents which are normally available from reduced GSH and nicotinamide nucleotides.

The physiological significance of an absence of extracellular Ca^{2+} has been supported by liver perfusion studies (Reed, 1992; Okuda et al., 1992). It is interesting to speculate upon situations in which such patho-physiological phenomena may be applicable. For example, a subject of considerable interest to researchers and clinicians is organ transplantation. The success rate of these procedures is highly subject to the length of preservation, and currently it is thought that cellular alterations occurring during preservation, such as pro-teolytic conversion of xanthine dehydrogenase to its oxidase form, leads to enhanced oxygen radical generation that contributes to tissue necrosis. As preservation is generally conducted in Ca^{2+}-free fluid, it is conceivable that changes in intracellular Ca^{2+} status set the stage for a massive production of active oxygen species upon reperfusion. Temporary deprivation of Ca^{2+} to perfused organs, followed by reperfusion with Ca^{2+}-containing medium, leads to an influx of Ca^{2+} even under anoxic conditions. Earlier data do not indicate a major role for oxygen-derived free radicals in the damage caused by Ca^{2+} deprivation. Today the evidence is overwhelming that oxygen-derived free radicals have an important role in experimentally induced reperfusion injury. These radicals are thought to cause significant mitochondrial dysfunction following reoxygenation of anoxic cells (Malis and Bonventre, 1988; Vlessis and Mela-Riker, 1989; Ambrosio and Chiariello, 1991; Darley-Usmar et al., 1991). The Ca^{2+} omission model of hepatocyte injury provides ample evidence that oxidative stress is a major component of cell injury during Ca^{2+} deprivation and that such stress is closely associated with mitochondrial functions. A major question is the possible role of permeability transition of the mitochondrial inner membrane in the loss of mitochondrial functions. With recent studies the role of mitochondrial processes in the maintenance of cell viability under pathological conditions is being understood better.

VIII. MITOCHONDRIAL GSH

Several studies have shown that mitochondrial GSH functions as a discrete pool separate from cytosolic GSH. A report by Jocelyn (1975) demonstrated that mitochondrial GSH is impermeable to the inner membrane following isolation of mitochondria, and Wahlländer et al. (1979) reported the concentration of mitochondrial GSH (10 mM) is higher than cytosolic GSH (7 mM). Studies by Meredith and Reed (1982) demonstrated different rates of GSH turnover in the cytosol and mitochondria confirming the existence of separate intracellular GSH pools. The ratio of GSH:GSSG in mitochondria is approximately 10:1 under normal (untreated) conditions. Unlike cytosolic GSSG, mitochondrial GSSG is not effluxed from the soluble compartment as reported by Olafsdottir and Reed (1988). This study demonstrated that during oxidative stress induced with tert-butyl hydroperoxide, GSSG is accumulated in the mitochondrial matrix and eventually reduced back to GSH. However, as the redox state of the mitochondria increased, an increase in protein mixed disul-fides was also observed. This study concluded that mitochondria are more sensitive to redox changes in GSH:GSSG than the cytosol, and therefore mitochondria may be more susceptible to the damaging effects of oxidative stress. These findings suggest that under certain experimental conditions, irreversible cell injury due to oxidative challenge may result from irreversible changes in mitochondrial function.

Studies addressing mitochondria as target organelles of certain types of irreversible cell injury, as related to oxidative stress as well as disrupted Ca^{2+} homeostasis, represents an area of intense investigation (Nohl et al., 1989; Chacon and Acosta, 1991; Turrens et al., 1991; Cleeter et al., 1992; Schulze-Osthoff et al., 1992). GSH-dependent protection against lipid peroxidation has been demonstrated in mitochondria, nuclei, microsomes, and cytosol of rat liver. Lipid peroxidation induced in mitochondria is inhibited by respiratory substrates such as succinate, which leads indirectly to reduction of ubiquinone to ubiquinol. The latter is a potent antioxidant (Takayanagi et al., 1980; Bindoli et al., 1982; Mészaros et al., 1982). The essential factor in preventing accumulation of lipid peroxides and lysis of membranes in mitochondria, however, is glutathione peroxidase (Flohe and Schlegel, 1971). Although the prevention of free radical attack on membrane lipids may occur by an electron shuttle that utilizes vitamin E and GSH in microsomes, similar activity may not be capable of inhibiting peroxidation in mitochondria (Reddy et al., 1982; McCay et al., 1988). Instead, mitochondrial GSH transferase(s) may prevent lipid peroxidation in mitochondria by a non-selenium glutathione-dependent peroxidase activity. Three GSH transferases have been isolated from the mitochondrial matrix (Kraus, 1980), and nearly 5% of the mitochondrial outer membrane protein consists of microsomal glutathione transferase (Morgenstern et al., 1984). GSH transferase in the outer mitochondrial membrane could provide the GSH-dependent protection of mitochondria by scavenging lipid radicals by a mechanism that requires vitamin E and is abolished by bromosulfophthalein (Morgenstern et al., 1984).

Since there is growing experimental evidence that oxidative stress is a factor in the neuropathology of several adult neurodegenerative disorders, many studies are being directed toward understanding the oxidative stress associated with these diseases (Coyle and Puttfarcken, 1993). For example, Adams et al. (1993) have examined brain oxidative stress induced by tert-butyl hydroperoxide and have concluded from studies with 2- and 8-month-old mice that aging makes the brain more susceptible to oxidative damage.

The pathogenesis of cell injury and death during oxidative stress may be caused by calcium ion-induced permeability transition of the mitochondrial inner membrane. A feature of oxidative stress-induced cell injury is morphological and functional changes in mitochondria (Takeyama et al., 1993; Carini et al., 1992; Masaki et al., 1989). Mitochondrial dysfunction, a critical event in anoxia/reoxygenation injury of liver sinusoidal endothelial cells, appears due to the permeability transition since acidosis protects against lethal oxidative injury (Bronk and Gores, 1991) as does a combination of cyclosporin A plus trifluoperazine (Fujii et al., 1994).

IX. MITOCHONDRIAL
PERMEABILITY TRANSITION

Several decades of studies with isolated mitochondria from a variety of tissues indicate that under certain experimental conditions, the mitochondrial inner membrane, which is normally impermeable to solutes, becomes permeable (Malamed and Recknagel, 1959; Lehninger, 1962; Gunter and Pfeiffer, 1990). Although the mechanism(s) by which this occurs remains controversial, this process is frequently referred to as mitochondrial permeability transition (Gunter and Pfeiffer, 1990; Szabö and Zoratti, 1992). The permeability transition is characterized as a nonspecific and as a specific (Richter et al., 1990) Ca^{2+}-dependent inner membrane pore which presents itself in mitochondria following treatment with Ca^{2+} and a second agent, termed an inducing agent. Many inducing agents have been identified and vary greatly in both structure and function (Gunter and Pfeiffer, 1990). It is thought that these agents, in the presence of Ca^{2+}, act through a common mechanism during the permeability transition. Examples of inducing agents include inorganic phosphate, cytosolic factor, fatty acids, heavy metals, organic sulfhydryl reagents, and oxidants such as tert-butyl hydroperoxide (Gunter and Pfeiffer, 1990).

Several identifiable events of the permeability transition have been characterized in the last several years. These include inner membrane permeability to small ions and solutes with molecular weights <1200 Da., large amplitude swelling, loss of coupled functions, and sensitivity to cyclosporin A (CyA) (Al-Nasser and Crompton, 1986; Petronilli et al., 1993; Crompton et al., 1988; Broekemeier et al., 1989). In addition, loss of matrix proteins via the inner membrane pore has also been reported (Igbavboa et al., 1989). During permeability transition, Ca^{2+} as well as other ions, are rapidly released from the mitochondrial matrix, presumably via the inner membrane pore. Following inner membrane permeability and the release of matrix solutes, a colloidal osmotic pressure arises in the mitochondrial matrix due to the high concentration of proteins which are slow to equilibrate (Gunter and Pfeiffer, 1990). In order to correct the osmotic imbalance, the entrance of H_2O results in massive swelling of the mitochondria (Gunter and Pfeiffer, 1990).

Mitochondrial swelling under these conditions is termed large amplitude swelling. Although secondary to inner membrane permeability and solute release, large amplitude swelling occurs within a short time (3 to 10 min) and is easily detected by monitoring light scatter of mitochondrial suspensions at 540 nm. Monitoring mitochondrial swelling is a simple assay and is often utilized as an indicator of permeability transition. Different experimental conditions result, however, in different patterns and times of swelling, some with shorter or longer lag periods (Broekemeier et al., 1989).

Studies by Fournier et al. (1987) showed with isolated mitochondria that CyA treatment promoted retention of accumulated Ca^{2+}. Subsequent studies by

Crompton et al. (1988) were designed to examine the effect of CyA on pore opening since the loss of Ca^{2+} observed by Fournier et al. (1987) was possibly due to pore opening. This study revealed that CyA is a potent inhibitor of the permeability transition when the inducing agents inorganic phosphate or tert-butyl hydroperoxide were added in the presence of Ca^{2+}. Broekemeier et al. (1989) tested CyA as an inhibitor of the inner membrane pore in the presence of several different inducing agents and found that very low concentrations of CyA protected against permeability transition. These studies demonstrated the potency of CyA in preventing the permeability transition, as inhibition of inner membrane permeability is observed with concentrations as low as 100 pmoles CyA/mg mitochondrial protein.

Studies by Halestrap and Davidson (1990) have focused on the mechanism of pore formation and have proposed that the adenine nucleotide translocase is the putative pore structure. They hypothesize that the "m" conformation of this inner membrane protein is modified to the "c" conformation by the binding of negative effectors such as Ca^{2+}, cyclophilin, and inducing agents resulting in a protein conformation which can no longer function as an adenine nucleotide translocase but as a nonspecific pore. Atractyloside and bongkrekic acid are known modifiers of the translocase and have been helpful in proposing this mechanism. This study also demonstrates that both ATP and ADP are positive effectors of the putative permeability transition pore.

Richter et al. (1990) propose that pore formation is not required for Ca^{2+} release and suggest that ADP-ribosylation of inner membrane proteins is the responsible mechanism of permeability transition. They hypothesize that up-take of Ca^{2+} and the consequent recycling of Ca^{2+} results in NADPH oxidation followed by the hydrolysis of NAD+ forming a nicotinamide and an ADP-ribose moiety. Subsequent ADP-ribosylation of critical proteins results in inner membrane permeability allowing the specific release of Ca^{2+}.

The present understanding of permeability transition is that the pore structure is an allosteric inner membrane protein, perhaps the adenine nucleotide translocase, with several different regulatory sites which are effected by Ca^{2+}, Pi, oxidants, sulfhydryl reagents, heavy metals, cyclophilin, ADP, ATP, atractyloside, and bongkrekic acid. Depending upon which sites are modified may determine whether permeability transition occurs. It is hypothesized that inhibition of permeability transition with CyA results from the binding of cyclophilin, a matrix protein *cis-trans* isomerase thought to be involved with protein folding (Griffiths and Halestrap, 1991). Whether the process of permeability transition is physiological remains an intriguing question.

Several methods are available to detect/measure mitochondrial permeability transition in isolated mitochondria. Crompton and Costi (1988) have developed a method which exploits two aspects of the permeability transition: (1) the entry of radiolabeled sucrose into the mitochondrial matrix, a normally impermeable solute to the mitochondrial inner membrane and (2) the reversible nature of the pore. By adding a chelator of Ca^{2+}, such as EDTA, or CyA, a potent pore inhibitor, the inner membrane nonspecific pore can be closed. This allows for the entrapment of sucrose into the matrix which can then be quantified.

Release of added Ca^{2+} or other ions from mitochondria and its inhibition by CyA are also indicative of permeability transition (Gunter and Pfeiffer, 1990; Broekemeier et al., 1989). As previously discussed, loss of absorbance at 540 nm of mitochondrial suspensions is frequently used as an indicator of permeability transition in isolated mitochondria. However, since large amplitude swelling is a secondary event which does not always accompany permeability transition (Savage and Reed, 1994a), monitoring of solute movement may be a more sensitive parameter of permeability changes.

Mitochondrial GSH (307 Da) release may be a useful and sensitive endogenous indicator of permeability transition under the conditions we have examined (Savage et al., 1991; Savage and Reed, 1994a). Studies from our laboratory with rat liver mitochondria have shown that during permeability transition induced with 70 μM Ca^{2+} and 3 mM Pi, mitochondrial GSH is rapidly and completely released from the matrix and recovered extramitochondrially as GSH (Savage et al., 1991). This release is completely prevented by addition of 0.5 μM CyA, indicating that release occurs via the Ca^{2+}-dependent, CyA-sensitive inner membrane pore (Savage et al., 1991). GSH release under these conditions parallels Ca^{2+} release in that both of these molecules are nearly completely released during a 5 min incubation. Mitochondrial swelling, a secondary process, is somewhat slower, however, and this process results in a substantial loss of mitochondrial density which is prevented by the addition of CyA.

An interesting finding is that 3 mM Pi alone induces a permeability transition by a CyA-sensitive mechanism. Mitochondria undergo large amplitude swelling with Pi alone, although not as extensive as in mitochondria treated with both Ca^{2+} and Pi. However, complete GSH release occurs with Pi treatment alone, although at a slower rate than with the addition of both Ca^{2+} and Pi. This release and swelling appears due to the presence of 6–10 nmoles Ca^{2+}/ mg protein in the mitochondrial preparations. The effect of Pi is abolished following the addition of EGTA (data not shown). Although monitoring of swelling indicates that permeability transition is occurring, the finding suggests permeability transition is not as extensive relative to Ca^{2+} and Pi treatment. However, without exogenously added Ca^{2+}, monitoring of Ca^{2+} release is not a sensitive indicator of permeability transition, as endogenous Ca^{2+} levels are very low and therefore a significant change in Ca^{2+} levels is not observed.

Although swelling is detected when Pi is added alone, it is less than maximal swelling detected by light scatter in the presence of 70 μM Ca^{2+} and 3 mM Pi. However, electron microscopy of samples treated with Pi alone or Pi plus Ca^{2+} reveals an indistinguishable level of large amplitude swelling, with varied times (data not shown). We conclude from this information, combined with our biochemical data, that monitoring of Ca^{2+} release or swelling is not always a sensitive indicator of the degree of permeability transition. Monitor-

ing release of a highly concentrated endogenous solute, such as GSH release
by a CyA-sensitive mechanism, may be a sensitive indicator of inner membrane permeability transition.

More recent studies have supported this finding (Savage and Reed, 1994a).
Mitochondria treated with Ca^{2+} and Pi in the presence of metabolic inhibitors of
electron transport or ATP synthesis do not undergo mitochondrial large amplitude swelling as detected by light scatter and electron microscopy. However,
monitoring of matrix GSH indicates that GSH is released from the matrix into
the extramitochondrial environment. Treatment with 0.5 μM CyA prevents this
release of GSH, suggesting that release occurred via the Ca^{2+}-dependent inner
membrane pore. The rate and extent of GSH release varies with the different
conditions, however, release of this endogenous molecule may be a useful
marker of mitochondrial permeability transition.

X. OXIDATION OF PYRIDINE NUCLEOTIDES
DURING PERMEABILITY TRANSITION

Studies by Richter et al. (1990) propose that oxidation of pyridine nucleotides and the subsequent hydrolysis is the mechanism responsible for Ca^{2+}
release. These studies treated mitochondria with the oxidant, tert-butyl hydroperoxide, as the inducing agent in the presence of Ca^{2+}. Measurement of total
pyridine nucleotides in our model, in the presence of Pi rather than an oxidant,
revealed that pyridine nucleotides undergo extensive oxidation very early
during permeability transition, suggesting the involvement of endogenous
oxidative stress (Savage and Reed, 1994b). The loss of reduced pyridine
nucleotides was accounted for as the oxidized product, indicating that hydrolysis of oxidized pyridine nucleotides does not occur to a significant degree and
therefore does not appear to be involved with Ca^{2+} release under these conditions. In these experiments, GSSG did not exceed untreated control values,
which was approximately 0.5 nmoles GSSG/mg mitochondrial protein.

Conditions of permeability transition greatly diminished mitochondrial energy
status as indicated by the redox ratios of NAD+/NADH, NADP+/NADPH, and
the energy charge. Treatment with 0.5 μM CyA maintains pyridine nucleotide
ratios very similar to untreated controls while increasing slightly the energy
charge (Savage and Reed, 1994b). Oxidation of pyridine nucleotides in the
presence of Ca^{2+} and Pi suggest oxidative stress may be involved in this model
of permeability transition. Although increased levels of GSSG were not detected, possibly due to a very active glutathione redox cycle, or lipid peroxidation
in this system, oxidation of pyridine nucleotides are consistent with an oxidative stress. This is the first indication permeability transition may be intricately
related with oxidative stress. If this type of injury occurs at the cellular level,
irreversible cellular injury may occur if mitochondria cannot repair this damage. The physiological consequences of permeability transition and possible
relationship to oxygen toxicity are currently under investigation.

ACKNOWLEDGMENTS

Support in part for this work was by United States Public Health Service Grants ES01978, ES00210, ES0040, and HL49182.

REFERENCES

Adams, J. D. Jr., Wang, B., Klaidman, L. K., LeBel, C. P., Odunze, I. N., and Shah, D., New aspects of brain oxidative stress induced by *tert*-butylhydroperoxide, *Free Rad. Biol. Med.*, 15, 195, 1993.

Akasaka, S., Inactivation of transforming activity of plasmid DNA by lipid peroxidation, *Biochim. Biophys. Acta*, 867, 201, 1986.

Alin, P., Danielson, U. H., and Mannervik, B., 4-Hydroxyalk-2-enals are substrates for glutathione transferase, *FEBS Lett.*, 179, 267, 1985.

Al-Nasser, I. and Crompton, M., The reversible Ca2+-induced permeabilization of rat liver mitochondria, *Biochem. J.*, 239, 19, 1986.

Ambrosio, G. and Chiariello, M., Myocardial reperfusion injury: Mechanisms and management — A review, *Am. J. Med.*, 91, 86S, 1991.

Aust, S. D., Morehouse, L. A., and Thomas, C. E., Role of metals in oxygen radical reactions, *J. Free Radic. Biol. Med.*, 1, 3, 1985.

Babior, B. M., Oxidant from phagocytes: Agents of defense and destruction, *Blood*, 64, 959, 1984.

Babior, B. M., The respiratory burst oxidase, *Trends Biol. Sci.*, 12, 241, 1987.

Baker, S. P. and Hemsworth, B. A., Effect of mitochondrial lipid peroxidation on monoamine oxidase, *Biochem. Pharmacol.*, 27, 805, 1978.

Barclay, L. R. C., Bailey, A. M. H., and Kong, D., The antioxidant activity of α-tocopherol-bovine serum albumin complex in micellar and liposome autoxidations, *J. Biol. Chem.*, 260, 15809, 1985.

Bascetta, E., Gunstone, F. D., and Walton, J. C., Electron spin resonance study of the role of vitamin E and vitamin C in the inhibition of fatty acid oxidation in a model membrane, *Chem. Phys. Lipids*, 33, 207, 1983.

Bindoli, A., Cavallini, L., and Jocelyn, P., Mitochondrial lipid peroxidation by cumene hydroperoxide and its prevention by succinate, *Biochim. Biophys. Acta*, 681, 496, 1982.

Bindoli, A., Lipid peroxidation in mitochondria, *Free Rad. Biol. Med.*, 5, 247, 1988.

Boveris, A. and Cadenas, E., Production of superoxide in mitochondria, in *Superoxide Dismutase, II*, Oberley, L. W., Ed., CRC Press, Boca Raton, FL, 1982, 15.

Brambilla, G., Sciaba, L., Faggin, P., Maura, A., Marinari, U. M., Ferro, M., and Esterbauer, H., Cytotoxicity, DNA fragmentation and sister-chromatid exchange in Chinese hamster ovary cells exposed to the lipid peroxidation product 4-hydroxynonenal and homologous aldehydes, *Mutat. Res.*, 171, 169, 1986.

Brawn, K. and Fridovich, I., DNA strand scission by enzymically generated oxygen radicals, *Arch. Biochem. Biophys.*, 206, 414, 1981.

Broekemeier, K. M., Dempsey, M. E., and Pfeiffer, D. R., Cyclosporin A is a potent inhibitor of the inner membrane permeability transition in liver mitochondria, *J. Biol. Chem.*, 264, 7826, 1989.

Bronk, S. F. and Gores, G. J., Acidosis protects against lethal oxidative injury of liver sinusoidal endothelial cells, *Hepatology*, 14, 150, 1991.

Bucher, J. R., Tien, M., Morehouse, L. A., and Aust, S. D., Redox cycling and lipid peroxidation: The central role of iron chelates, *Fundam. Appl. Toxicol.*, 3, 222, 1983.

Burton, G. W., Joyce, A., and Ingold, K. U., Is vitamin E the only lipid-soluble, chain-breaking antioxidant in human blood plasma and erythrocyte membranes? *Arch. Biochem. Biophys.*, 221, 281, 1983.

Cadenas, E., Biochemistry of oxygen toxicity, *Annu. Rev. Biochem.*, 58, 79, 1989.

Carini, R., Parola, M., Dianzani, M. U., and Albano, E., Mitochondrial damage and its role in causing hepatocyte injury during stimulation of lipid peroxidation by iron nitriloacetate, *Arch. Biochem. Biophys.*, 297, 110, 1992.

Chacon, E. and Acosta, D., Mitochondrial regulation of superoxide by Ca2+: An alternate mechanism for the cardiotoxicity of doxorubicin, *Toxicol. Appl. Pharmacol.*, 107, 117, 1991.

Chance, B., Sies, H., and Boveris, A., Hydroperoxide metabolism in mammalian organs, *Physiol. Rev.*, 59, 527, 1979.

Christophersen, B. O., Formation of monohydroxy-polyenic fatty acids from lipid peroxides by a glutathione peroxidase, *Biochim. Biophys. Acta*, 164, 35, 1968.

Chu, F. F., Doroshow, J. H., and Esworthy, R. S., Expression, characterization, and tissue distribution of a new cellular selenium-dependent glutathione peroxidase, GSHPx-GI, *J. Biol. Chem.*, 268, 2571, 1993.

Cleeter, M. W., Cooper, J. M., and Schapira, A. H., Irreversible inhibition of mitochondrial complex I by 1-methyl-4-phenylpyridinium: Evidence for free radical involvement, *J. Neurochem.*, 58, 786, 1992.

Cowan, D. B., Weisel, R. D., Williams, W. G., and Mickle, D. A. G., Identification of oxygen responsive elements in the 5'-flanking region of the human glutathione peroxidase gene, *J. Biol. Chem.*, 268, 26904, 1993.

Coyle, J. T. and Puttfarcken, P., Oxidative stress, glutamate, and neuro-degenerative disorders, *Science*, 262, 689, 1993.

Crompton, M. and Costi, A., Kinetic evidence for a heart mitochondrial pore activated by Ca2+, inorganic phosphate and oxidative stress. A potential mechanism for mitochondrial dysfunction during cellular Ca2+ overload, *Eur. J. Biochem.*, 178, 489, 1988.

Crompton, M., Ellinger, H., and Costi, A., Inhibition by cyclosporin A of a Ca2+-dependent pore in heart mitochondria activated by inorganic phosphate and oxidative stress, *Biochem. J.*, 255, 357, 1988.

Dabeva, M. D., Petrov, P. T., Stoykova, A. S., and Hadjiolov, A. A., Contamination of detergent purified rat liver nuclei by cytoplasmic ribosomes, *Exp. Cell Res.*, 108, 467, 1977.

Darley-Usmar, V. M., Stone, D., Smith, D., and Martin, J. F., Mitochondria, oxygen and reperfusion damage, *Ann. Med.*, 23, 583, 1991.

Dean, R. T. and Cheeseman, K. H., Vitamin E protects against free radical damage in lipid environments, *Biochem. Biophys. Res. Commun.*, 148, 1277, 1987.

Di Mascio, P., Murphy, M. E., and Sies, H., Antioxidant defense systems: The role of carotenoids, tocopherols, and thiols, *Am. J. Clin. Nutr.*, 53, 194S, 1991.

Dillard, C. J., Kunert, K. J., and Tappel, A. L., Effects of vitamin E, ascorbic acid and mannitol on alloxan-induced lipid peroxidation in rats, *Arch. Biochem. Biophys.*, 216, 204, 1982.

Diplock, A. T., Antioxidant nutrients and disease prevention: An overview, *Am. J. Clin. Nutr.*, 53, 189S, 1991.

Dix, T. A. and Aikens, J., Mechanisms and biological relevance of lipid peroxidation initiation, *Chem. Res. Toxicol.* 6, 2, 1993.

Doba, T., Burton, G. W., and Ingold, K. U., Antioxidant and co-antioxidant activity of vitamin C. The effect of vitamin C, either alone or in the presence of vitamin E or a water-soluble vitamin E analogue, upon the peroxidation of aqueous multi-lamellar phospholipid liposomes, *Biochim. Biophys. Acta*, 835, 298, 1985.

Eichenberger, K., Bohni, P., Winterhalter, K. H., Kawato, S., and Richter, C., Microsomal lipid peroxidation causes an increase in the order of the membrane lipid domain, *FEBS Lett.*, 142, 59, 1982.

Fariss, M. W., Olafsdottir, K., and Reed, D. J., Extracellular calcium protects isolated rat hepatocytes from injury, *Biochem. Biophys. Res. Commun.*, 121, 102, 1984.

Flohe, L. and Schlegel, W., Glutathione peroxidase. IV. Intracellular distribution of the glutathione peroxidase system in the rat liver, *Hoppe-Seyler's Z. Physiol. Chem.*, 352, 1401, 1971.

Forman, H. J. and Boveris, A., Superoxide radical and hydrogen peroxide in mitochondria, in *Free Radicals in Biology,* Vol. V, Pryor, W. A., Ed., Academic Press, New York, 1982.

Fournier, N., Ducet, G., and Crevat, A., Action of cyclosporine on mitochondrial calcium fluxes, *J. Bioenerg. Biomembr.*, 19, 297, 1987.

Franke, W. W., Structure, biochemistry, and functions of the nuclear envelope, *Int. Rev. Cytol.*, Suppl. 4, 71, 1974.

Frenkel, K., Carcinogen-mediated oxidant formation and oxidative DNA damage, *Pharmac. Ther.*, 53, 127, 1992.

Fridovich, I., Superoxide dismutases, *Adv. Enzymol. Relat. Areas Mol. Biol.*, 41, 35, 1974.

Fujii, Y., Johnson, M. E., and Gores, G. J., Mitochondrial dysfunction during anoxia/reoxygenation injury of liver endothelial cells, *Hepatology*, 20, 177, 1994.

Fukuzawa, K., Takase, S., and Tsukatani, H., The effect of concentration on the antioxidant effectiveness of α-tocopherol in lipid peroxidation induced by superoxide free radicals, *Arch. Biochem. Biophys.*, 240, 117, 1985.

Goldstein, S., Meyerstein, D., and Czapski, G., The Fenton reagents, *Free Rad. Biol. Med.*, 15, 435, 1993.

Golumbic, C. and Mattill, H. A., Antioxidants and the autoxidation of fats. XIII. The antioxygenic action of ascorbic acid in association with tocopherols, hydroquinones and related compounds, *J. Am. Chem. Soc.*, 63, 1279, 1941.

Griffiths, E. J. and Halestrap, A. P., Further evidence that cyclosporin A protects mitochondria from calcium overload by inhibiting a matrix peptidyl-prolyl cis-trans isomerase. Implications for the immunosuppressive and toxic effects of cyclosporin, *Biochem. J.,* 274, 611, 1991.

Gruger, E. H. and Tappel, A. L., Reactions of biological antioxidants: III. Composition of biological membranes, *Lipids*, 6, 147, 1971.

Gunter, T. E. and Pfeiffer, D. R., Mechanisms by which mitochondria transport calcium, *Am. J. Physiol.*, 258, C755, 1990.

Haenen, G. R. M. M. and Bast, A., Protection against lipid peroxidation by a microsomal glutathione-dependent labile factor, *FEBS Lett.*, 159, 24, 1983.

Halestrap, A. P. and Davidson, A. M., Inhibition of Ca2(+)-induced large-amplitude swelling of liver and heart mitochondria by cyclosporin is probably caused by the inhibitor binding to mitochondrial-matrix peptidyl-prolyl cis-trans isomerase and preventing it interacting with the adenine nucleotide translocase, *Biochem. J.*, 268, 153, 1990.

Halliwell, B. and Gutteridge, J. M. C., Role of free radicals and catalytic metal ions in human disease: An overview, in *Methods in Enzymology*, Packer, L. and Glazer, A. N., Eds., Academic Press, San Diego, vol. 186(B), 1990, 1.

Harlan, J. M., Callahan, K. S., Schwartz, B. R., and Harker, L. A., Glutathione redox cycle protects cultured endothelial cells against lysis by extracellularly generated hydrogen peroxide, *J. Clin. Invest.*, 73, 706, 1984.

Herschberger, L. A. and Tappel, A. L., Effect of vitamin E on pentane exhaled by rats treated with methyl ethyl ketone peroxide, *Lipids*, 17, 686, 1982.

Higashi, T., Tateishi, N., and Naruse, A. et al., Novel physiological role of liver glutathione as a reservoir of L-cysteine, *J. Biochem.*, 82, 117, 1977.

Hill, K. E. and Burk, R. F., Influence of vitamin E and selenium on glutathione-dependent protection against microsomal lipid peroxidation, *Biochem. Pharmacol.*, 33, 1065, 1984.

Hjelle, J. J. and Petersen, D. R., Metabolism of malondialdehyde by rat liver aldehyde dehydrogenase, *Toxicol. Appl. Pharmacol.*, 70, 57, 1983.

Horrobin, D. F., Is the main problem in free radical damage caused by radiation, oxygen, and other toxins, the loss of membrane essential fatty acids, rather than the accumulation of toxic materials? *Med. Hypotheses*, 35, 23, 1991.

Igbavboa, U., Zwizinski, C. W., and Pfeiffer, D. R., Release of mitochondrial matrix proteins through a Ca2+-requiring, cyclosporin-sensitive pathway, *Biochem. Biophys. Res. Commun.*, 161, 619, 1989.

Ito, A., Hayashi, S., and Yoshida, T., Participation of a cytochrome b5-like hemoprotein of outer mitochondrial membrane (OM cytochrome b) in NADH-semidehydroascorbic acid reductase activity of rat liver, *Biochem. Biophys. Res. Commun.*, 101, 591, 1981.

Jocelyn, P. C., Some properties of mitochondrial glutathione, *Biochim. Biophys. Acta.*, 396, 427, 1975.

Jones, D. P., Eklow, L., Thor, H., and Orrenius, S., Metabolism of hydrogen peroxide in isolated hepatocytes: Relative contributions of catalase and glutathione peroxidase in decomposition of endogenously generated H2O2, *Arch. Biochem. Biophys.*, 210, 505, 1981.

Keberle, H., The biochemistry of desferrioxamine and its relation to iron metabolism, *Ann. NY Acad. Sci.*, 119, 758, 1964.

Kornbrust, D. J. and Mavis, R. D., Relative susceptibility of microsomes from lung, heart, liver, kidney, brain and testes to lipid peroxidation: Correlation with vitamin E content, *Lipids*, 15, 315, 1980.

Kosower, N. S. and Kosower, E., Glutathione status of cells, *Int. Rev. Cytol.*, 54, 289, 1978.

Kraus, P., Resolution, purification and some properties of three glutathione transferases from rat liver mitochondria, *Hoppe-Seyler's Z. Physiol. Chem.*, 361, 9, 1980.

Lehninger, A. L., Water uptake and extrusion by mitochondria in relation to oxidative phosphorylation, *Phys. Rev.*, 42, 467, 1962.

Leung, H. W., Vang, M. J., and Mavis, R. D., The cooperative interaction between vitamin E and vitamin C in suppression of peroxidation of membrane phospholipids, *Biochim. Biophys. Acta*, 664, 266, 1981.

Liebler, D. C., Kaysen, K. L, and Kennedy, T. A., Redox cycles of vitamin E: Hydrolysis and ascorbic acid dependent reduction of 8a-(alkyldioxy)tocopherones, *Biochemistry*, 28, 9772, 1989.

Liebler, D. C., Kling, D. S., and Reed, D. J., Antioxidant protection of phospholipid bilayers by α-tocopherol, *J. Bio. Chem.*, 261, 12114, 1986.

Litov, R. E., Matthews, L. C., and Tappel, A. L., Vitamin E protection against in vivo lipid peroxidation initiated in rats by methyl ethyl ketone peroxide as monitored by pentane, *Toxicol. Appl. Pharmacol.*, 59, 96, 1981.

Machlin, L., Ed., *Vitamin E: A Comprehensive Treatise*, Marcel Dekker, New York, 1980.

Maguire, J. J., Wilson, D. S., and Packer, L., Mitochondrial electron transport-linked tocopheroxyl radical reduction, *J. Biol. Chem.*, 264, 21462, 1989.

Malamed, S. and Recknagel, R. O., The osmotic behavior of the sucrose-inaccessible space of mitochondrial pellets from rat liver, *J. Biol. Chem.* 234, 3027, 1959.

Malis, C. D. and Bonventre, J. V., Mechanism of calcium potentiation of oxygen free radical injury to renal mitochondria, *J. Biol. Chem.*, 261, 14201, 1986.

Malis, C. D. and Bonventre, J. V., Susceptibility of mitochondrial membranes to calcium and reactive oxygen species: Implications for ischemic and toxic tissue damage, *Prog. Clin. Biol. Res.*, 282, 235, 1988.

Masaki, N., Kyle, M. E., Serroni, A., and Farber, J. L., Mitochondrial damage as a mechanism of cell injury in the killing of cultured hepatocytes by tert-butyl hydroperoxide, *Arch. Biochem. Biophys.*, 270, 672, 1989.

McCay, P. B., Vitamin E: Interactions with free radicals and ascorbate, *Ann. Rev. Nutr.*, 5, 323, 1985.

McCay, P. B., Brueggemann, G., Lai, E. K., and Powell, S. R., Evidence that α-tocopherol functions cyclically to quench free radicals in hepatic microsomes. Requirement for glutathione and a heat-labile factor, *Ann. N.Y. Acad. Sci.*, 570, 32, 1989.

McCay, P. B., Gibson, D. D., Fong, K. L., et al., Effect of glutathione peroxidase activity on lipid peroxidation in biological membranes, *Biochim. Biophys.*, 261, 1, 1988.

McCay, P. B., Lai, E. K., Powell, S. R., et al., Vitamin E functions as an electron shuttle for glutathione-dependent "free radical reductase" activity in biological membrane, *Fed. Proc. Fed. Am. Soc. Exp. Biol.*, 45, 1729, 1976.

McCay, P. B., Poyer, J. L., Pfeifer, P. M., May, H. E., and Gilliam, J. M., A function for α-tocopherol: Stabilization of the microsomal membrane from radical attack during TPNH-dependent oxidations, *Lipids*, 6, 297, 1971.

Meredith, M. J. and Reed, D. J., Status of the mitochondrial pool of glutathione in the isolated hepatocyte, *J. Biol. Chem.*, 257, 3747, 1982.

Meredith, M. J. and Reed, D. J., Depletion *in vitro* of mitochondrial glutathione in rat hepatocytes and enhancement of lipid peroxidation by adriamycin and 1,3-bis(2-chloroethyl)-1-nitrosourea (BCNU), *Biochem. Pharmacol.*, 32, 1382, 1983.

Mészaros, L., Tihanyi, K., and Horvath, I., Mitochondrial substrate oxidation-dependent protection against lipid peroxidation, *Biochim. Biophys. Acta*, 731, 675, 1982.

Miller, D. M., Buettner, G. R., and Aust, S. D., Transition metals as catalysts of "autoxidation" reactions, *Free Rad. Biol. Med.*, 8, 95, 1990.

Misra, H. P., Generation of superoxide free radical during the autoxidation of thiols, *J. Biol. Chem.*, 249, 2151, 1974.

Morgenstern, R., Lundqvist, G., and Andersson, G., et al., The distribution of microsomal glutathione transferase among different organelles, different organs, and different organisms, *Biochem. Pharmacol.*, 33, 3609, 1984.

Mukai, F. H. and Goldstein, B. D., Mutagenicity of malonaldehyde, a decomposition product of peroxidized polyunsaturated fatty acids, *Science*, 191, 868, 1976.

Muller, A., Cadenas, E., and Graf, P., et al., A novel biologically active seleno-organic compound — I. Glutathione peroxidase-like activity in vitro and antioxidant capacity of PZ 51 (Ebselen), *Biochem. Pharmacol.*, 33, 3235, 1984.

Muller, D. P. R., Lloyd, J. K., and Wolff, O. H., in *Biology of Vitamin E (Ciba Foundation Symposium 101)*, Pitman Books, London, 1983, 106.

Murphy, M. E., Scholich, H., Wefers, H., and Sies, H., Alpha-tocopherol in microsomal lipid peroxidation, *Ann. N. Y. Acad. Sci.*, 570, 480, 1989.

Nakamura, W., Hosoda, S., and Hayashi, K., *Biochim. Biophys. Acta*, 358, 251, 1974.

Neubert, D., Wojtszak, A. B., and Lehninger, A. L., Purification and enzymatic identity of mitochondrial contraction factors I and II, *Proc. Natl. Acad. Sci., USA*, 48, 1651, 1962.

Niki, E., Kawakami, A., Yamamoto, Y., and Kamiya, Y., Oxidation of lipids. VIII. Synergistic inhibition of oxidation of phosphatidylcholine liposome in aqueous dispersion by vitamin E and vitamin C, *Bull. Chem. Soc. Jpn.*, 58, 1971, 1985.

Niki, E., Saito, T., and Kamiya, Y., The role of vitamin C as an antioxidant, *Chem. Lett.*, 631, 1983.

Niki, E., Yamamoto, Y., Komuro, E., and Sato, K., Membrane damage due to lipid oxidation, *Am. J. Clin. Nutr.*, 53, 201S, 1991.

Nohl, H., deSilva, D., and Summer, K. H., 2,3,7,8, tetrachlorodibenzo-p-dioxin induces oxygen activation associated with cell respiration, *Free Rad. Biol. Med.*, 6, 369, 1989.

O'Brien, P. J. and Hawco, F. J., Hydroxyl-radical formation during prostaglandin formation catalyzed by prostaglandin cyclo-oxygenase, *Biochem. Soc. Trans.*, 6, 1169, 1978.

Okuda, M., Hsien-Chang, Le., Chance, B., and Kumar, C., Depletion and repletion of Ca^{2+} in the perfused rat liver, *J. Lab. Clin. Med.*, 120, 57, 1992.

Olafsdottir, K. and Reed, D. J., Retention of oxidized glutathione by isolated rat liver mitochondria during hydroperoxide treatment, *Biochim. Biophys. Acta*, 964, 377, 1988.

Packer, J. E., Slater, T. F., and Willson, R. L., Direct observation of a free radical interaction between vitamin E and vitamin C, *Nature (London)*, 278, 737, 1979.

Paraidathathu, T., de Groot, H., and Kehrer, J. P., Production of reactive oxygen by mitochondria from normoxid and hypoxic rat heart tissue, *Free Rad. Biol. Med.*, 13, 289, 1992.

Pascoe, G. A. and Reed, D. J., Relationship between cellular calcium and vitamin E metabolism during protection against cell injury, *Arch. Biochem. Biophys.*, 253, 287, 1987.

Petronilli, V., Cola, C., and Bernardi, P., Modulation of the mitochondrial cyclosporin A-sensitive permeability transition pore. II. The minimal requirements for pore induction underscore a key role for transmembrane electrical potential, matrix pH, and matrix Ca^{2+}, *J. Biol. Chem.*, 268, 1011, 1993.

Plaa, G. L. and Witschi, H., Chemicals, drugs, and lipid peroxidation, *Annu. Rev. Pharmacol. Toxicol.*, 16, 125, 1976.

Porter, N. A., Chemistry of lipid peroxidation, *Methods Enzymol.*, 105, 273, 1984.

Pryor, W. A., Oxy-radicals and related species: Their formation, lifetimes, and reactions, *Annu. Rev. Physiol.*, 48, 657, 1986.

Pryor, W. A. and Godber, S. S., Noninvasive measures of oxidative stress status in humans, *Free Rad. Biol. Med.*, 10, 177, 1991.

Reddy, C. C., Sholz, W. W., and Thomas, C. B., et al., Vitamin E-dependent reduced glutathione inhibition of rat liver microsomal lipid peroxidation, *Life Sci.*, 31, 571, 1982.

Reed, D. J., Cellular defense mechanisms against reactive metabolites, in *Bioactivation of Foreign Compounds*, Anders, M. W., Ed., Academic Press, Orlando, FL, 1985, 71.

Reed, D. J., Regulation of reductive processes by glutathione, *Biochem. Pharmacol.*, 35, 7, 1986.

Reed, D. J., Glutathione: Toxicological implications, *Annu. Rev. Pharmacol Toxicol.*, 30, 603, 1990.

Reed, D. J., Interaction of vitamin E, ascorbic acid, and glutathione in protection against oxidative stress, in *Vitamin E in Health and Disease*, Packer, L., Ed., Marcel Dekker, New York, Chap. 21, 1992, 269.

Reed, D. J. and Beatty, P. W., Biosynthesis and regulation of glutathione. Toxicological implications, in *Reviews in Biochemical Toxicology,* Hodgson, E., Bend, J. R., and Philpot, R. M., Eds., Elsevier Press, New York, 1980, 213.

Reed, D. J. and Ellis, W. W., Influence of γ-glutamyl transpeptidase inactivation on the status of extracellular glutathione and glutathione conjugates, in *Biological Reactive Intermediates, IIA*, Snyder, R., Parke, C. V., Kocsis, J. J., et al., Eds., Plenum Press, New York, 1982, 75.

Retsky, K. L., Freeman, M. W., and Frei, B., Ascorbic acid oxidation product(s) protect human low density lipoprotein against atherogenic modification, *J. Biol. Chem.*, 268, 1304, 1993.

Rice-Evans, C. A. and Diplock, A. T., Current status of antioxidant therapy, *Free Rad. Biol. Med.*, 15, 77, 1993.

Richter, C., Do mitochondrial DNA fragments promote cancer and aging? *FEBS Lett.*, 241, 1, 1988.

Richter, C., Theus, M., and Schlegel, J., Cyclosporine A inhibits mitochondrial pyridine nucleotide hydrolysis and calcium release, *Biochem. Pharmacol.*, 40, 779, 1990.

Roots, R. and Okada, S., Estimation of life times and diffusion distances of radicals involved in x-ray-induced DNA strand breaks of killing of mammalian cells, *Radiat. Res.*, 64, 306, 1975.

Rowley, D. A. and Halliwell, B., Superoxide-dependent formation of hydroxyl radicals in the presence of thiol compounds, *FEBS Lett.*, 138, 33, 1982.

Sandy, M. S., Moldéus, P., and Ross, D., et al., Cytotoxicity of the redox cycling compound diquat in isolated hepatocytes: Involvement of hydrogen peroxide and transition metals, *Arch. Biochem. Biophys.*, 259, 29, 1987.

Savage, M. K., Jones, D. P., and Reed, D. J., Calcium- and phosphate-dependent release and loading of glutathione by liver mitochondria, *Arch. Biochem. Biophys.*, 290, 51, 1991.

Savage, M. K. and Reed, D. J., Release of mitochondrial glutathione and calcium by a cyclosporin A-sensitive mechanism occurs without large amplitude swelling, *Arch. Biochem. Biophys.*, 315, 142, 1994a.

Savage, M. K. and Reed, D. J., Oxidation of pyridine nucleotides and depletion of ATP and ADP during calcium- and inorganic phosphate-induced mitochondrial permeability transition, *Biochem. Biophys. Res. Commun.*, 200, 1615, 1994b.

Scarpa, M., Rigo, A., Maiorino, M., Ursini, F., and Gregolin, C., Formation of α-tocopherol radical and recycling of α-tocopherol by ascorbate during peroxidation of phosphatidylcholine liposomes. An electron paramagnetic resonance study, *Biochim. Biophys. Acta*, 801, 215, 1984.

Schwartz, C. J., Valente, A. J., Sprague, E. A., Kelley, J. L., and Nerem, R. M., The pathogenesis of atherosclerosis: An overview, *Clin. Cardiol.*, 14, 1, 1991.

Schulze-Osthoff, K., Bakker, A. C., Vanhaesebroeck, B., Beyaert, R., Jacob, W. A., and Fiers, W., Cytotoxic activity of tumor necrosis factor is mediated by early damage of mitochondrial functions. Evidence for the involvement of mitochondrial radical generation, *J. Biol. Chem.*, 267, 5317, 1992.

Sevanian, A., Hacker, A. D., and Elsayed, N., Influence of vitamin E and nitrogen dioxide on lipid peroxidation in rat lung and liver microsomes, *Lipids*, 17, 269, 1982.

Shigenaga, M. K. and Ames, B. N., Assays for 8-hydroxy-2'-deoxyguanosine: A biomarker of in vivo oxidative DNA damage, *Free Radic. Biol. Med.*, 10, 211, 1991.

Sies, H., Introduction, in *Oxidative Stress*, Sies, H., Ed., Academic Press, Orlando, 1985, 1.

Silva, J. M., McGirr, L., and O'Brien, P. J., Prevention of nitrofurantoin-induced cytotoxicity in isolated hepatocytes by fructose, *Arch. Biochem. Biophys.*, 289, 313, 1991.

Sohal, R. S., Aging, cytochrome oxidase activity, and hydrogen peroxide release by mitochondria, *Free Rad. Biol. Med.*, 14, 583, 1993.

Steinberg, D., Parthasarathy, S., Carew, T. E., Khoo, J. C., and Witztum, J. L., Beyond cholesterol. Modifications of low-density lipoprotein that increase its atherogenicity, *N. Engl. J. Med.*, 320, 915, 1989.

Svingen, B. A., Buege, J. A., O'Neal, F. O., and Aust, S. D., The mechanism of NADPH-dependent lipid peroxidation. The propagation of lipid peroxidation, *J. Biol. Chem.*, 254, 5892, 1979.

Svingen, B. A., O'Neal, F. O., and Aust, S. D., The role of superoxide and singlet oxygen in lipid peroxidation, *Photochem. Photobiol.*, 28, 803, 1978.

Szabö, I. and Zoratti, M., The mitochondrial megachannel is the permeability transition pore, *J. Bioenerg. Biomembr.*, 24, 111, 1992.

Takayanagi, R., Takeshige, K., and Minakami, S., NADH- and NADPH-dependent lipid peroxidation in bovine heart submitochondrial particles. Dependence on the rate of electron flow in the respiratory chain and an antioxidant role of ubiquinol, *Biochem. J.*, 192, 853, 1980.

Takeyama, N., Matsuo, N., and Tanaka, T., Oxidative damage to mitochondria is mediated by the Ca^{2+}-dependent inner membrane permeability transition, *Biochem. J.*, 294, 719, 1993.

Tappel, A. L., Vitamin E and free radical peroxidation of lipids, *Ann. N.Y. Acad. Sci.*, 203, 12, 1972.

Tappel, A. L., Lipid peroxidation damage to cell components, *Fed. Proc.*, 32, 1870, 1973.

Tappel, A. L., Brown, W. D., Zalkin, H., and Maier, V. P., Unsaturated lipid peroxidation catalyzed by hematin compounds and its inhibition by vitamin E., *J. Am. Oil Chem. Soc.*, 38, 5, 1961.

Tateishi, N., Higashi, T., and Naruse, A. et al., Rat-liver glutathione. Possible role as a reservoir of cysteine, *J. Nutr.*, 107, 51, 1977.

Thomas, C. E. and Reed, D. J., Effect of extracellular Ca^{++} omission on isolated hepatocytes. I. Induction of oxidative stress and cell injury, *J. Pharmacol. Exp. Ther.*, 245, 493, 1988a.

Thomas, C. E. and Reed, D. J., Effect of extracellular Ca^{++} omission on isolated hepatocytes. II. Loss of mitochondrial membrane potential and protection by inhibitors of uniport Ca^{++} transduction, *J. Pharmacol. Exp. Ther.*, 245, 501, 1988b.

Thomas, C. E. and Reed, D. J., Radical-induced inactivation of kidney Na^+, K^+-ATPase: Sensitivity to membrane lipid peroxidation and the protective effect of vitamin E, *Arch. Biochem. Biophys.*, 281, 96, 1990.

Tien, M., Bucher, J. R., and Aust, S. D., Thiol-dependent lipid peroxidation, *Biochem. Biophys. Res. Commun.*, 107, 279, 1982.

Tirmenstein, M. A. and Reed, D. J., Characterization of glutathione-dependent inhibition of lipid peroxidation of isolated rat liver nuclei, *Arch. Biochem. Biophys.*, 261, 1, 1988.

Tirmenstein, M. A. and Reed, D. J., Effects of glutathione on the α-tocopherol-dependent inhibition of nuclear lipid peroxidation, *J. Lipid Res.*, 30, 959, 1989.

Tribble, D. L., Aw, T. Y., and Jones, D. P., The pathophysiological significance of lipid peroxidation in oxidative cell injury, *Hepatology*, 7, 377, 1987.

Turrens, J. F., Beconi, M., Barilla, J., Chavez, U. B., and McCord, J. M., Mitochondrial generation of oxygen radicals during reoxygenation of ischemic tissues, *Free Rad. Res. Commun.*, 12-13, 681, 1991.

Turrens, J. F. and McCord, J. M., Mitochondrial generation of reactive oxygen species, in *Free Radicals, Lipoproteins, and Membrane Lipids*, Paulet, A. C., Douste-Blazy, L., and Paoletti, R., Eds., Plenum Press, New York, 1990, 203.

Ueda, K., Kobayashi, S., Morita, J., and Komano, T., Site-specific DNA damage caused by lipid peroxidation products, *Biochim. Biophys. Acta*, 824, 341, 1985.

Ursini, F., Maiorino, M., and Gregolin, C., The selenoenzyme phospholipid hydroperoxide glutathione peroxidase, *Biochim. Biophys. Acta*, 839, 62, 1985.

Vignais, P. M. and Vignais, P. V., Fuscin, an inhibitor of mitochondrial SH-dependent transport-linked functions, *Biochim. Biophys. Acta*, 325, 357, 1973.

Vlessis, A. A. and Mela-Riker, L., Mitochondrial calcium metabolism during tissue reperfusion, *Prog. Clin. Biol. Res.*, 299, 63, 1989.

Wahlländer, A., Sobell, S., Sies, H., Linke, I., and Müller, M., Hepatic mitochondrial and cytosolic glutathione content and the subcellular distribution of GSH-S-transferases, *FEBS Lett.*, 97, 138, 1979.

Weitzel, F. and Wendel, A., Selenoenzymes regulate the activity of leukocyte 5-lipoxygenase via the peroxide tone, *J. Biol. Chem.*, 268, 6288, 1992.

White, I. N. H., Role of liver glutathione in acute toxicity of retrorsine to rats, *Chem. Biol. Interact.*, 13, 333, 1976.

Willson, R. L., in *Biology of Vitamin E (Ciba Foundation Symposium 101)*, Pitman Books, London, 1983, 19.

Womack, M., Kremmer, K. S., and Rose, W. C., The relation of cysteine and methionine to growth, *J. Biol. Chem.*, 121, 403, 1937.

Chapter 4

AFLATOXINS

Margaret M. Manson

CONTENTS

I. Introduction 69

II. AFB$_1$ Metabolism 70
   A. Metabolism in Rodents 70
   B. Metabolism in Humans 73

III. Species Susceptibility 75

IV. Mechanisms of Resistance 75

V. Biomonitoring Human Exposure 76

VI. Oncogenes and Tumor Suppressor Genes 77

VII. Summary 78

References 79

I. INTRODUCTION

Scientific interest in the natural toxins produced by the genus of mold named *Aspergillus* was triggered in 1960 when the condition known as "Turkey X" disease was first described (reviewed in Heathcote and Hibbert, 1978). The death of tens of thousands of turkeys, ducklings, and pheasants prompted research which led to the identification of aflatoxins in the fungus *A. flavus* as the toxic metabolites responsible for the disease. Aflatoxin was first isolated from Brazilian peanut meal (Sargeant et al., 1961), but can contaminate a wide variety of cereals and other crops, given the right conditions of humidity and temperature. Although mostly a problem in countries around the equator and in the southern hemisphere, as recently as 1988 widespread contamination of corn in the American Midwest was reported (Ezzell, 1988), and since food stuffs are imported from countries where contamination is a recurring problem, these compounds are of worldwide concern. Aflatoxins can also be found further up the food chain, for example in cows' milk (Iongh et al., 1964), and in some African countries they have also been identified in human milk (Wild et al., 1987; Lamplugh et al., 1988). It is now widely recognized that fungal contamination of foodstuffs can cause mycotoxicosis and induce cancer in both animals and man.

As well as its negative impact worldwide as a contributing factor to human disease and to crop spoilage, on the positive side, aflatoxin has been invaluable as a tool in the elucidation of mechanisms of toxicity and carcinogenesis. The MRC Toxicology Unit has been closely involved in a number of aspects of this research over many years, some of which are illustrated below.

II. AFB$_1$ METABOLISM

A. METABOLISM IN RODENTS

Data on the toxicity and carcinogenicity of the various aflatoxins (B$_1$, B$_2$, G$_1$, G$_2$, M$_1$) and their metabolites have been collated in a recent IARC (International Agency for Research on Cancer) monograph (1993). Of all the aflatoxins studied, aflatoxin B$_1$ (AFB$_1$) is the most acutely toxic and the most potent carcinogen, so that much of the research has been concentrated on this molecule. The adverse effects are well documented in rat liver where biliary hyperplasia and hepatocyte necrosis occur during the acute phase, while preneoplastic altered foci of hepatocytes and hyperplastic nodules precede tumor formation (Newberne and Butler, 1969). Morphologically distinct tumor types occur, including trabecular hepatocellular carcinoma, pseudoglandular tumors, and more rarely cholangiocarcinoma (Butler et al., 1969).

In order to exert its toxicity AFB$_1$ requires metabolism via the cytochrome P450 monooxygenase pathway. A number of P450 isozymes are involved, and the proportions of resulting metabolites are dependent on the isozyme profile of the target tissue. In rodent liver the major intermediate of aflatoxin is AFB$_1$-8,9-epoxide (Figure 4.1), which happens to be the most potent mutagenic and carcinogenic metabolite. In rat this is predominantly generated by P450 2C11 (Neal and Judah, unpublished observations; Ishii et al., 1986). Recent evidence suggests that both an endo and exo epoxide can result from cytochrome P450 activation (Raney et al., 1992). The highly reactive exo epoxide, thought to be the ultimate carcinogen, interacts with the N7 position of guanine in DNA, which, if not repaired, can lead to carcinogenesis (Essigman et al., 1977; Lin et al., 1977; Martin and Garner, 1977).

However, cellular consequences of aflatoxin toxicity, as also found in the toxicity of other compounds which are converted into reactive metabolites, are not only determined by phase I metabolism, but also by phase II detoxifying enzyme systems. Most significant of these in rodents are the glutathione S-transferases (GSTs), of which the alpha class appears to be the most active at conjugating AFB$_1$-epoxide to glutathione (Neal and Green, 1983; Coles et al., 1985). Following conjugation, the next step in the pathway toward mercapturic acid formation is the removal of the glutamate moiety by gamma glutamyl

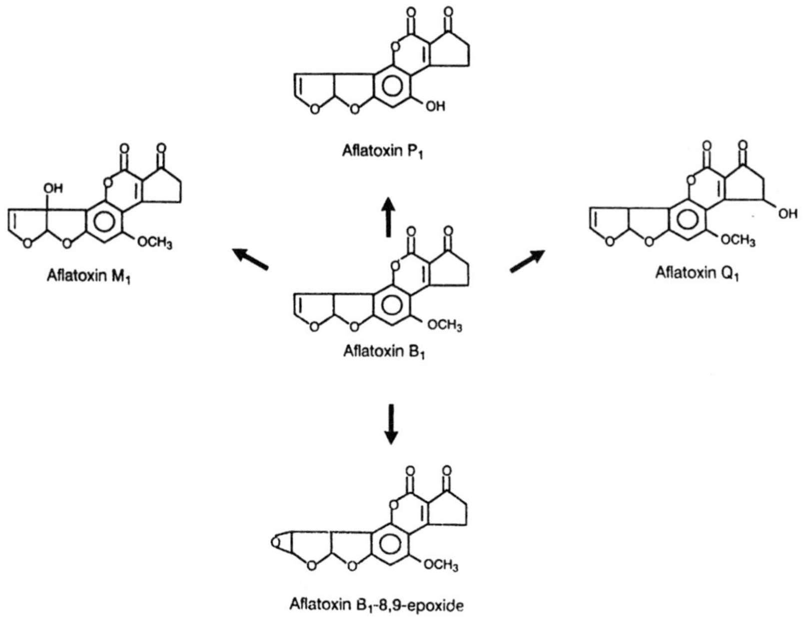

FIGURE 4.1. Major metabolites of aflatoxin B$_1$.

transpeptidase (GGT). This enzyme is also responsible for recycling extracellular glutathione (Reed and Ellis, 1981), particularly in the kidney. We were able to show that GGT is involved *in vivo* in the breakdown of aflatoxin-glutathione conjugates by comparing biliary excretion of AFB$_1$ metabolites in male and female rats and by studying the effect of AT125 (L-[αS,5S] α-amino-3-chloro-4,5-dihydro-5-isoxazoleacetic acid), an inhibitor of GGT activity (Moss et al., 1984). First, when comparing control male and female Fischer rats, females were found to contain twice as much GGT in their livers. While in normal adult male rats GGT is present only in bile duct cells, in females some periportal hepatocytes also contain the activity (Manson and Smith, 1984). After a single injection of AFB$_1$, the amount of the cysteinyl-glycine conjugate in bile was significantly higher in females. Second, in male rats, which had been fed with AFB$_1$ long enough to induce altered foci of hepatocytes which were positive for GGT, a sevenfold increase in enzyme activity led to a tenfold increase in the amount of AFB$_1$-cysteinylglycine present in the bile. When GGT activity was inhibited by treatment with AT125, then no AFB$_1$-cysteinylglycine was formed and the intact GSH conjugate was excreted into the bile. Each of the steps in the pathway from the formation of the aflatoxin epoxide-GSH conjugate to AFB$_1$-N-acetyl cysteine (Figure 4.2) has been characterized by biochemical and analytical means, including NMR and mass spectrometry (Moss et al., 1983, 1985).

An alternative pathway for the epoxide is hydrolysis either spontaneously or possibly through the action of epoxide hydrolase to form the AFB$_1$-dihydrodiol. This can then interact with proteins, for example through the ε-amino group of lysine, which in turn may lead to cytotoxicity (Neal et al., 1981; Sabbioni et al., 1987, 1990). An alternative pathway for the dihydrodiol, recently elucidated, appears to be the formation of an aflatoxin dialcohol via the action of an aldehyde reductase (Hayes et al., 1993a, Judah et al., 1993). This novel metabolite was detected in livers of rats which had been treated with the antioxidant ethoxyquin (EQ), but not in livers of control, adult animals. The dialcohol results from reduction of the dialdehyde form of AFB$_1$-dihydrodiol. Since the dialdehyde reacts readily with proteins and thereby may cause

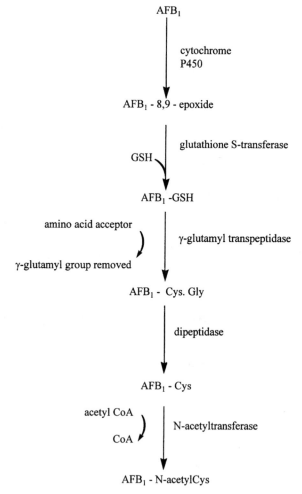

FIGURE 4.2.   Proposed metabolism of AFB$_1$ by formation of the AFB$_1$-GSH conjugate and its disposal via the mercapturic acid pathway (after Moss et al., 1985).

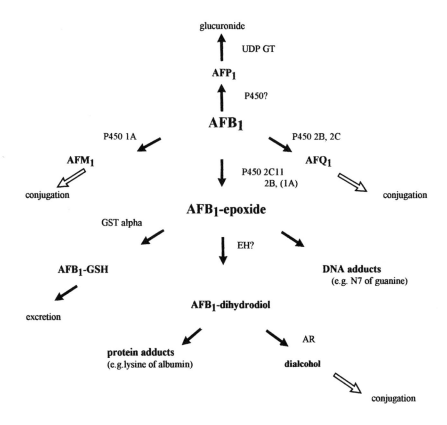

FIGURE 4.3. Rodent metabolism of AFB_1 (EH = epoxide hydratase, UDP-GT = UDP glucuronyl transferase, AR = aldehyde reductase, GSH = glutathione). Open arrows indicate steps in the pathway not yet fully defined.

cytotoxicity, it has been suggested that its reduction to the dialcohol may be another detoxification route for AFB_1 (Judah et al., 1993). The enzyme which catalyses the formation of AFB_1-dialcohol is a member of the aldo-keto reductase supergene family (Hayes et al., 1993b). The current knowledge on rodent metabolism of AFB_1 is summarized in Figure 4.3.

B. METABOLISM IN HUMANS

There is a good deal of circumstantial evidence implicating AFB_1 in human liver cancer (IARC, 1993). For example, the liver cancer rates per 100,000 population per year for several regions of the world correlated well with the estimated intake of AFB_1 contaminated diet (Linsell and Peers, 1977). In addition, episodes of acute toxicity in humans associated with consumption of AFB_1 contaminated diets have also been reported. For example in India, aflatoxicosis in the form of jaundice, fever, ascites, edema of the feet, and vomiting occurred after consumption of heavily contaminated maize. Men appeared more susceptible than women and 106 people died (Krishnamachari et al., 1975). Of those who survived, follow-up studies at 3 and 5 years showed almost complete recovery from acute poisoning (Tandon and Tandon, 1989). Exposure to aflatoxins has been linked to the etiology of kwashiorkor (Hendrickse and Maxwell, 1989). Several groups have studied the primary metabolism of AFB_1 by human liver microsomes (Moss and Neal, 1985; Shimada and Guengerich, 1989; Aoyama et al., 1990; Ramsdell and Eaton, 1990; Forrester et al., 1990; Kirby et al., 1993). As with rodents, multiple P450 isozymes are involved, but in humans the major metabolite is AFQ_1 instead of the epoxide, and the P450 3A family appears to be principally involved. As would be expected in the human population there is considerable interindividual variation in metabolism (Forrester et al., 1990). Unlike some rodent species, humans show rather poor ability to conjugate the substantial amounts of epoxide that are formed (Moss and Neal, 1985), but where this does occur the mµ class of GSTs appears to be predominantly involved (Liu et al., 1991). Human metabolism of AFB_1 is summarized in Figure 4.4.

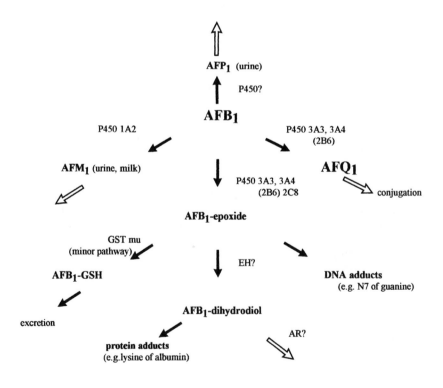

FIGURE 4.4.   Human metabolism of AFB_1.

III. SPECIES SUSCEPTIBILITY

As with most xenobiotics, there is a wide range of susceptibility to both the toxicity and carcinogenicity of AFB_1 among different species. This appears to be determined not only by the phase I metabolism, but very much by the ability for conjugation via the phase II enzymes, particularly the GSTs (O'Brien et al., 1983; Neal et al., 1987; Hayes et al., 1991a). At one end of the scale, avian species are extremely sensitive to the cytotoxicity of aflatoxins. They appear to be very efficient at metabolizing the AFB_1 to the epoxide, but have a poor capacity for conjugation with glutathione. In contrast, mice, while less effective at producing the epoxide, are much more efficient at conjugation, which helps to explain their relative resistance to the compound. The different glutathione-S transferase enzymes also have varying abilities for conjugation of AFB_1 metabolites, and one of the isozymes expressed in mouse liver is particularly effective (Degen and Neumann, 1981; O'Brien et al., 1983; Neal et al., 1987; Hayes et al., 1991a). Monroe and Eaton (1988) showed that if glutathione was depleted in mouse liver by buthionine S-sulphoxamine or diethyl maleate, then the sensitivity of this species toward AFB_1 was greatly enhanced. Judging from the information available from *in vitro* studies of human metabolism, namely that a significant amount of epoxide is produced and GSH conjugation is rather poor, man should rank somewhere close to rats in terms of sensitivity. However, many factors are involved in determining overall sensitivity of an individual, including sex, age, and exposure to other xenobiotics (Hayes et al., 1991a).

IV. MECHANISMS OF RESISTANCE

During the carcinogenic process induced in liver, the population of cells within the altered hepatic foci becomes resistant to the toxic effects of the inducing carcinogen (Solt and Farber, 1976; Judah et al., 1977). When such foci induced by AFB_1 in the rat are examined immunocytochemically in serial sections, the following changes can be observed: (a) loss of cytochromes P450, particularly 2C11; (b) an increase in GSTs, particularly the pi class GST 7-7 (Green et al., 1993) and alpha class GSTs (Harrison et al., 1990); and (c) an increase in GGT (Manson, 1983; Green et al., 1993). Thus the cells within the foci are less able to metabolize the AFB_1 to the epoxide, but at the same time are better able to conjugate any epoxide which is formed with glutathione and dispose of it via the mercapturic acid pathway.

A situation similar to that described above pertains when animals are treated with certain chemopreventive agents prior to receiving a carcinogen. In this case the changes in drug metabolizing enzymes within the hepatocytes protect the animal from both the toxic and carcinogenic effects. A particular case which we have studied in detail is the prevention of AFB_1 carcinogenesis in rat liver by pretreatment with the antioxidant EQ (Cabral and Neal, 1983; Manson et al., 1987). Once again an alteration in the isozyme profile of cytochrome P450 was responsible for altering the metabolism of AFB_1, so that the less toxic, less carcinogenic metabolites AFM_1 and AFQ_1 were predominantly formed. In addition, GSTs were elevated, some three- to fourfold as measured by the conjugation of a the model substrate, chlorodinitrobenzene (CDNB), but in the order of 100-fold with respect to AFB_1 conjugation. GGT activity was also greatly elevated, but with a predominantly periportal distribution, in contrast to the focal distribution seen during carcinogenesis (Mandel et al., 1987). As a result of these changes AFB_1 binding to DNA was inhibited by more than 97%. A study by Kensler et al. (1986), using the same strain of rat and the same chemicals, reached very similar conclusions.

The discrepancy between the increase in total GST activity induced in rat liver by EQ as measured by CDNB, compared to the much larger increase in AFB_1 conjugating activity suggested that perhaps a minor isozyme of GST was involved in AFB_1 conjugation. Competition studies using antibodies against the different classes of GST showed that only antisera against the alpha class were efficient at knocking out the AFB_1 conjugating activity (Hayes et al., 1991a). Further investigation revealed that a rat isozyme of GST with extremely basic characteristics was involved, and this showed close homology to a constitutive mouse alpha isoform. This particular form, called Yc_2 in the rat, is present in fetal liver, is virtually absent from control adult liver, but is very effectively induced by EQ and is also present in the preneoplastic altered foci of hepatocytes which are induced by AFB_1 (Hayes et al., 1991a, b, 1993a). This isozyme is also effectively induced by other xenobiotics such as phenobarbital, butylated hydroxyanisole and butylated hydroxytoluene (McLellan et al., 1994).

V. BIOMONITORING HUMAN EXPOSURE

Aflatoxins have long been suspected to be human toxins and hepatic carcinogens, but no direct study was feasible until relatively recently, when assays to measure an individual's exposure to aflatoxin became available (Groopman et al., 1984, 1985; Autrup et al., 1987; Hsieh et al., 1988; Gan et al., 1988; Wild et al., 1990). Estimates of human intake had previously been made by relating the aflatoxin content of food stuffs to the amount of food consumed; these not only proved to be highly inaccurate, but could not take into account interindividual variations in metabolism (see for example, Peers and Linsell, 1973; Ngindu et al., 1982). A number of techniques are now available to monitor AFB_1 metabolites in human samples, including immunoassay (Wild et al., 1987), and immunoaffinity purification and HPLC (Groopman, 1985; Sabbioni et al., 1990). In a recent study Ross et al. (1992) assayed for urinary AFB_1, AFP_1, and AFM_1 and also for DNA adducts to assess the relationship between aflatoxin exposure and liver cancer in China. They found that subjects with liver cancer were more likely than controls to have detectable concentrations of any of the metabolites, but that the highest risk was associated with AFP_1. It is unfortunate that in this study it was not possible to measure levels of AFQ_1, since, as already mentioned, this appears to be the major human metabolite produced in *in vitro* microsomal incubations. A criticism of this method is that the urinary excretion of aflatoxin metabolites and adducts is very rapid, thus being indicative only of exposure in the preceding 24 hours, and it has been pointed out that an assay for aflatoxin-serum albumin adducts would provide a measure of longer-term exposure (Ross et al., 1992; Hall and Wild, 1992). In the same study Ross et al. went on to provide some of the most convincing evidence to date for a synergistic effect between exposure to AFB_1 and infection with the hepatitis B virus (HBV). This is still an area of controversy since in parts of the world where liver cancer is common, there is usually a high risk of food being contaminated with AFB_1 and also a high carrier rate of HBV. The relative contribution of each of these two factors to hepatocellular carcinoma has so far been difficult to establish.

VI. ONCOGENES AND TUMOR SUPPRESSOR GENES

The discovery of oncogene activation as a contributing, if not the single, factor responsible for transformation in many tumor types posed the question as to whether such genes might be involved in AFB_1 hepatocarcinogenesis. Studies of rat tumors implicated activation of the ras oncogene in a sizable proportion of tissue samples, when these were analyzed by transfection of NIH3T3 mouse fibroblast cells with tumor DNA. Evidence for activation of both Ki and N ras was obtained (McMahon et al., 1986, 1987; Sinha et al., 1988; Soman and Wogan, 1993). The predominant mutation in the Ki-ras gene was a G.C to A.T base transition at the second nucleotide in codon 12. At present it is not known how late an event ras activation is in these tumors, or whether it occurs in any of the preneoplastic lesions.

Mutations in the tumor suppressor gene p53 are one of the most common genetic changes in human cancer. The majority of such mutations cause both a loss of tumor suppressor activity and a gain of oncogenic function through abnormal expression of genes normally regulated by p53. Transgenic mice with a mutant or a deleted p53 gene exhibit an increased susceptibility to cancer (Lavigueur et al., 1989; Donehower et al., 1992). However, unlike the ras oncogene where mutations are found in only three codons, many different mutations have been identified in the p53 gene. In 1991 two papers were published which implicated the tumor suppressor gene p53 in human hepatocarcinogenesis (Hsu et al., 1991; Bressac et al., 1991). The interesting fact about these two studies, one of liver patients from China and the other from South Africa, was that both showed a hot spot for mutation in codon 249 of the p53 gene. And what was even more interesting was that the G to T transversion was consistent with the formation of an AFB_1 adduct at this site. This naturally generated a flurry of activity to confirm the finding in further human studies and also to produce an animal model in order to prove that AFB_1 was in fact involved. Subsequent clinical studies have produced conflicting data. In one report, in which liver tumor tissues from 107 geographically and ethnically diverse sources were examined, the mutation rate in p53 observed in tumors from regions of high AFB_1 exposure was 25% compared to only 12% in low exposure regions. This was lower than the 50% previously reported and only 2 out of 107 samples had a mutation in codon 249 (Buetow et al., 1992). In a Japanese study which set out to determine whether the mutation pattern in p53 could be used diagnostically as a marker for multiple hepatocellular carcinoma, the mutation rate was as high as 65%, although only 3 out of 17 cases had a mutation at codon 249 (Oda et al.,1992). A study of hepatocellular carcinoma in Britain, where exposure to AFB_1 would not be a risk factor, detected abnormalities in the p53 gene in 2 out of 19 cases, with no mutation at codon 249 (Challen et al., 1992). An immunohistochemical study of 58 white patients with hepatocellular carcinoma concluded that p53 mutation, or overexpression was a rare finding in patients exposed to low dietary aflatoxin (Laurent-Puig et al., 1992). Recently a study using the human cell line HepG2 exposed to AFB_1 showed mutations in codon 249 (Aguilar et al., 1993).

Attempts to find an animal model to demonstrate AFB_1 involvement in p53 mutation have so far met with little success. For example, in a study of nine tumors induced in nonhuman primates, only one was positive, and the mutation was a G to T transversion in codon 175 (Fujimoto et al., 1992).

It has been suggested, on the basis of the specific mutations caused in p53 by AFB_1 and by other carcinogens such as UV light and benzo-a-pyrene in cigarette smoke, that this molecule may offer an opportunity to identify particular carcinogens and help define biological mechanisms involved in development of human cancer. The frequency and type of p53 mutations would act as a molecular dosimeter of carcinogen exposure and so provide information on the molecular epidemiology of cancer risk in humans (Harris, 1993).

VII. SUMMARY

Despite the fact that AFB_1 contamination is not a major concern in the western world, it should be obvious from the examples illustrated above that its use as a model compound has provided valuable insight into a wide range of basic mechanisms in carcinogenesis. Many of these have wide-ranging applicability to other compounds, tissues, and species. Aflatoxin affects the regulation of many different genes, and understanding the mechanisms by which it can up- or down-regulate expression will provide a much greater understanding of how a cell responds to stress (be it chemical or oxidative) and how it becomes transformed, as well as helping to determine to what extent there is a coordinate response among the affected genes. There is still an abundant crop of information crucial to the understanding of carcinogenesis to be harvested from the study of this fascinating molecule.

REFERENCES

Aguilar, F., Hussain, S. P., and Cerutti, P., Aflatoxin B₁ induces a transversion of G→T in codon 249 of the p53 tumour suppressor gene in human hepatocytes, *Proc. Natl. Acad. Sci. U.S.A.*, 90, 8586, 1993.

Aoyama, T., Yamano, S., Guzelian, P. S., Gelboin, H. V., and Gonzalez, F. J., Five of 12 forms of vaccinia virus-expressed human hepatic cytochrome P450 metabolically activate aflatoxin B₁, *Proc. Natl. Acad. Sci. U.S.A.*, 87, 4790, 1990.

Autrup, H., Seremet, T., Wakhisi, J., and Wasunna, A., Aflatoxin exposure measured by urinary excretion of aflatoxin B₁-guanine adduct and hepatitis B virus infection in areas with different liver cancer incidence in Kenya, *Cancer Res.*, 47, 3430, 1987.

Bressac, B., Kew, M., Wands, J., and Ozturk, M., Selective G to T mutations of p53 gene in hepatocellular carcinoma from Southern Africa, *Nature (London)*, 350, 429, 1991.

Buetow, K. H., Sheffield, V. C., Zhu, M., Zhou, T., Shen, F., Hino, O., Smith, M., McMahon, B. J., Lanier, A. P., London, W. T., Redeker, A. G., and Govindarajan, S., Low frequency of p53 mutations observed in a diverse collection of primary hepatocellular carcinomas, *Proc. Natl. Acad. Sci. U.S.A.*, 89, 9622, 1992.

Butler, W. H., Greenblatt, M., and Lijinsky, W., Carcinogenesis in rats by aflatoxins B₁, G₁ and B₂, *Cancer Res.*, 29, 2206, 1969.

Cabral, J. R. P. and Neal, G. E., The inhibitory effects of ethoxyquin on the carcinogenic action of aflatoxin B₁ in rats, *Cancer Lett.*, 19, 125, 1983.

Challen, C., Lunec, J., Warren, W., Collier, J., and Bassendine, M. F., Analysis of the p53 tumour suppressor gene in hepatocellular carcinomas from Britain, *Hepatology*, 16, 1362, 1992.

Coles, B., Meyer, D. J., Ketterer, B., Stanton, C. A., and Garner, R. C., Studies on the detoxification of microsomally activated aflatoxin B₁ by glutathione and glutathione transferases *in vitro*, *Carcinogenesis*, 6, 693, 1985.

Degen, G. H. and Neumann, A. G., Differences in aflatoxin B₁-susceptibility of rat and mouse are correlated with the capability in vitro to inactivate aflatoxin B₁-epoxide, *Carcinogenesis*, 2, 299, 1981.

Donehower, L. A., Harvey, M., Slagle, B. L., McArthur, M. J., Montgomery, C. A., Butel, J. S., and Bradley, A. Mice deficient for p53 are developmentally normal but susceptible to spontaneous tumours, *Nature (London)*, 356, 215, 1992.

Essigman, J. M., Croy, R. G., Nadzan, A. M., Busby, W. F., Reinhold, V. N., Buchi, G., and Wogan, G. N., Structural identification of the major DNA adduct formed by aflatoxin B₁ *in vitro*, *Proc. Natl. Acad. Sci. U.S.A.*, 74, 1870, 1977.

Ezzell, C., Aflatoxin contamination of U.S. corn, *Nature (London)*, 335, 757, 1988.

Forrester, L. M., Neal, G. E., Judah, D. J., Glancey, M. J., and Wolf, C. R., Evidence for involvement of multiple forms of cytochrome P-450 in aflatoxin B₁ metabolism in human liver, *Proc. Natl. Acad. Sci. U.S.A.*, 87, 8306, 1990.

Fujimoto, Y., Hampton, L. L., Luo, L-di., Worth, P. J., and Thorgeirsson, S. S., Low frequency of p53 gene mutation in tumours induced by aflatoxin B₁ in non-human primates, *Cancer Res.*, 52, 1044, 1992.

Gan, L.-S., Skipper, P. L., Peng, X., Groopman, J., Chen, J.-S., Wogan, G. N., and Tannenbaum, S. R., Serum albumin adducts in the molecular epidemiology of aflatoxin carcinogenesis: correlation with aflatoxin B₁ intake and urinary excretion of aflatoxin M₁, *Carcinogenesis*, 9, 1323, 1988.

Green, J. A., Stamatoglou, S. C., Gordon, L. A., and Manson, M. M., Use of multiple markers to identify a subset of preneoplastic foci in liver most likely to progress to tumour, *Human Exp. Toxicol.*, 12, 576, 1993.

Groopman, J. D. Molecular dosimetry methods of assessing human aflatoxin exposures, in *The Toxicology of Aflatoxins: Human Health, Veterinary and Agricultural Significance*, Eaton, D. L. and Groopman, J. D., Eds., Academic Press, New York, 1994, 259.

Groopman, J. D., Donahue, P. R., Zhu, J., Chen, J., and Wogan, G. N., Aflatoxin metabolism in humans; detection of metabolites and nucleic acid adducts in urine by affinity chromatography, *Proc. Natl. Acad. Sci. U.S.A.*, 82, 6492, 1985.

Groopman, J. D., Trudel, L. J., Donahue, P. R., Marshat-Rothstein, A., and Wogan, G. N., High affinity monoclonal antibodies for aflatoxins and their application to solid phase immunoassays, *Proc. Natl. Acad. Sci. U.S.A.*, 81, 7728, 1984.

Hall, A. J. and Wild, C. P., Aflatoxin biomarkers, *Lancet*, 339, 1413, 1992.

Harris, C. C., p53: At the crossroads of molecular carcinogenesis and risk assessment, *Science*, 262, 1980, 1993.

Harrison, D. J., May, L., Hayes, J. D., and Neal, G. E., Glutathione S-transferase localization in aflatoxin B$_1$-treated rat livers, *Carcinogenesis*, 11, 927, 1990.

Hayes, J. D., Judah, D. J., Ellis, E. M., McLellan, L. I., and Neal, G. E., Role of glutathione S-transferase and aldehyde reductase in resistance to aflatoxin B$_1$, in *Structure and Function of Glutathione Transferases*, Tew, K. D., Pickett, C. B., Mantle, T. J., Mannervik, B., and Hayes, J. D., Eds., CRC Press, Boca Raton, FL, 1993a, 3.

Hayes, J. D., Judah, D. J., McLellan, L. I., Kerr, L. A., Peacock, S. D., and Neal, G. E., Ethoxyquin-induced resistance to aflatoxin B$_1$ in the rat is associated with the expression of a novel alpha class glutathione S-transferase subunit Yc$_2$, which possesses high catalytic activity for aflatoxin B$_1$-8,-9-epoxide, *Biochem. J.*, 279, 385, 1991b.

Hayes, J. D., Judah, D. J., McLellan, L. I., and Neal, G. E., Contribution of the glutathione S-transferases to the mechanisms of resistance to aflatoxin B$_1$, *Pharmacol. Ther.*, 50, 443, 1991a.

Hayes, J. D., Judah, D. J., and Neal, G. E., Resistance to aflatoxin B$_1$ is associated with the expression of a novel aldo-keto reductase which has catalytic activity towards a cytotoxic aldehyde containing metabolite of the toxin, *Cancer Res.*, 53, 3887, 1993b.

Hayes, J. D., Nguyen, T., Judah, D. J., Petersson, D. G., and Neal, G. E., Cloning of cDNAs from fetal rat liver encoding glutathione S-transferase Yc polypeptides, *J. Biol. Chem.*, 269, 1, 1994.

Heathcote, J. G. and Hibbert, J. R., Aflatoxins — chemical and biological aspects, in *Developments in Food Science 1*, Heathcote, J. G. and Hibbert, J. R., Eds., Elsevier, Amsterdam, 1978.

Hendrickse, R. G. and Maxwell, S. M., Aflatoxins and child health in the tropics, *J. Toxicol-Toxin Rev.*, 8, 31, 1989.

Hsieh, L.-L., Hsu, S.-W., Chen, D.-S., and Santella, R. M., Immunological detection of aflatoxin B$_1$-DNA adducts formed *in vivo*, *Cancer Res.*, 48, 6328, 1988.

Hsu, I. C., Metcalf, R. A., Sun, T., Welsh, J. A., Wang, N. J., and Harris, C. C., Mutational hotspot in the p53 gene in human hepatocellular carcinomas, *Nature (London)*, 350, 427, 1991.

International Agency for Research on Cancer, Some naturally occurring substances: Food items and constituents, heterocyclic aromatic amines and mycotoxins, *IARC Monograph*, 56, 245, 1993.

de Iongh, H., Vles, R. O., and van Pelt, J. G., Investigation of the milk of mammals fed an aflatoxin-containing diet, *Nature (London)*, 202, 466, 1964.

Ishii, K., Maeda, K., Kamakati, T., and Kato, R., Mutagenic activation of aflatoxin B$_1$ by several forms of purified cytochrome P450, *Mutation Res.*, 174, 85, 1986.

Judah, D. J., Hayes, J. D., Yang, J.-C., Lian, L.-Y., Roberts, G. C. K., Farmer, P. B., Lamb, J. H., and Neal, G. E., A novel aldehyde reductase with activity towards a metabolite of aflatoxin B$_1$ is expressed in rat liver during carcinogenesis and following the administration of an antioxidant, *Biochem. J.*, 292, 13, 1993.

Judah, D. J., Legg, R. F., and Neal, G. E., Development of resistance to cytotoxicity during aflatoxin carcinogenesis, *Nature*, 265, 343, 1977.

Kensler, T. W., Egner, P. A., Davidson, N. E., Roebuck, B. D., Pikul, A., and Groopman, J. D., Modulation of aflatoxin metabolism, aflatoxin-N^7-guanine formation and hepatic tumourigenesis in rats fed ethoxyquin: Role of induction of glutathione S-transferases, *Cancer Res.*, 46, 3924, 1986.

Kirby, G. M., Wolf, C. R., Neal, G. E., Judah, D. J., Henderson, C. J., Srivatanakul, P., and Wild, C. P., *In vitro* metabolism of aflatoxin by normal and tumourous liver tissue from Thailand, *Carcinogenesis*, 14, 2613, 1993.

Krishnamachari, K. A. V. R., Bhat, R. V., Nagarajan, V., and Tilak, T. B. G., Hepatitis due to aflatoxicosis. An outbreak in western India, *Lancet i*, 1061, 1975.

Lamplugh, S. M., Hendrickse, R. G., Apeagyei F., and Mwanmut, D. D., Aflatoxins in breast milk, neonatal cord blood, and serum of pregnant women, *Br. Med. J.*, 296, 968, 1988.

Laurent-Puig, P., Flejou, J. F., Fabre, M., Bedossa, P., Belghiti, J., Gayral, F., and Franco, D., Overexpression of p53: A rare event in a large series of white patients with hepatocellular carcinoma, *Hepatology*, 16, 1171, 1992.

Lavigueur, A., Maltby, V., Mock, D., Rossant, J., Pawson, T., and Bernstein, A., High incidence of lung, bone, and lymphoid tumours in transgenic mice overexpressing mutant alleles of the p53 oncogene, *Mol. Cell Biol.*, 9, 3982, 1989.

Lin, J. K., Miller, J. A., and Miller, E. C., 2,3-Dihydro-2-(gua-7-yl)-3-hydroxy-aflatoxin B_1 a major acid hydrolysis product of aflatoxin B_1-DNA or ribosomal RNA adducts formed in hepatic microsome-mediated reactions and in rat liver *in vivo*, *Cancer Res.*, 37, 4430, 1977.

Linsell, C. A. and Peers, F. G., Aflatoxin and liver cell cancer, *Trans. R. Soc. Trop. Med. Hyg.*, 71, 471, 1977.

Liu, Y. H., Taylor, J., Linko, P., Lucier, G. W., and Thompson, C., Glutathione S-transferase mu in human lymphocyte and liver. Role in modulating formation of carcinogen-derived DNA adducts, *Carcinogenesis*, 12, 2269, 1991.

Mandel, H. G., Manson, M. M., Judah, D. J., Simpson, J. L., Green, J. A., Forrester, L. M., Wolf, C. R., and Neal, G. E., Metabolic basis for the protective effect of the antioxidant ethoxyquin on aflatoxin B_1 hepatocarcinogenesis in the rat, *Cancer Res.*, 47, 5218, 1987.

Manson, M. M., Biphasic early changes in rat liver γ-glutamyl transpeptidase in response to aflatoxin B_1, *Carcinogenesis*, 4, 467, 1983.

Manson, M. M., Green, J. A., and Driver, H. E., Ethoxyquin alone induces preneoplastic changes in rat kidney whilst preventing induction of such lesions in liver by aflatoxin B_1, *Carcinogenesis*, 8, 723, 1987.

Manson, M. M. and Smith, A. G., Effect of hexachlorobenzene on male and female rat hepatic gamma-glutamyltranspeptidase levels, *Cancer Lett.*, 22, 227, 1984.

Martin, C. N. and Garner, R. C., Aflatoxin B_1-oxide generated by chemical or enzymic oxidation of aflatoxin B_1 causes guanine substitution in nucleic acids, *Nature (London)*, 267, 863, 1977.

McLellan, L. I., Judah, D. J., Neal, G. E., and Hayes, J. D., Regulation of aflatoxin B_1-metabolizing aldehyde reductase and glutathione S-transferase by chemoprotectors, *Biochem. J.*, 300, 117, 1994.

McMahon, G., Davies, E., and Wogan, G. N., Characterization of c-Ki-*ras* oncogene alleles by direct sequencing of enzymatically amplified DNA from carcinogen-induced tumours, *Proc. Natl. Acad. Sci. U.S.A.*, 84, 4974, 1987.

McMahon, G., Hanson, L., Lee, J.-J., and Wogan, G. N., Identification of an activated c-Ki-ras oncogene in rat liver tumours induced by aflatoxin B_1, *Proc. Natl. Acad. Sci. U.S.A.*, 83, 9418, 1986.

Monroe, D. H. and Eaton, D. L., Effects of modulation of hepatic glutathione on biotransformation and covalent binding of aflatoxin B_1 to DNA in the mouse, *Toxicol. Appl. Pharmacol.*, 94, 118, 1988.

Moss, E. J., Judah, D. J., Przybylski, M., and Neal, G. E., Some mass-spectral and n.m.r. analytical studies of a glutathione conjugate of aflatoxin B_1, *Biochem. J.*, 210, 227, 1983.

Moss, E. J., Manson, M. M., and Neal, G. E., Effect of manipulation of γ-glutamyl transpeptidase levels on biliary excretion of aflatoxin B_1 conjugates, *Carcinogenesis*, 5, 869, 1984.

Moss, E. J. and Neal, G. E., The metabolism of aflatoxin B_1 by human liver, *Biochem. Pharmacol.*, 34, 3193, 1985.

Moss, E. J., Neal, G. E., and Judah, D. J., The mercapturic acid pathway metabolites of a glutathione conjugate of aflatoxin B_1, *Chem-Biol. Interactions*, 55, 139, 1985.

Neal, G. E. and Green, J. A., The requirement for glutathione S-transferase in the conjugation of activated aflatoxin B_1 during aflatoxin hepatocarcinogenesis in the rat, *Chem. Biol. Interactions*, 45, 259, 1983.

Neal, G. E., Judah, D. J., Stirpe, F., and Patterson, D. S. P., The formation of 2,3-dihydroxy-2,3-dihydro aflatoxin B_1 by the metabolism of aflatoxin B_1 by liver microsomes isolated from certain avian and mammalian species and the possible role of this metabolite in the acute toxicity of aflatoxin B_1, *Toxicol. Appl. Pharmacol.*, 58, 431, 1981.

Neal, G. E., Nielsch, U., Judah, D. J., and Hulbert, P. B., Conjugation of model substrates or microsomally-activated aflatoxin B_1 with reduced glutathione, catalysed by cytosolic glutathione S-transferases in livers of rats, mice and guinea pigs, *Biochem. Pharmacol.*, 36, 4269, 1987.

Newberne, P. M. and Butler, W. H., Acute and chronic effects of aflatoxin on the liver of domestic and laboratory animals: A review, *Cancer Res.*, 29, 236, 1969.

Ngindu, A., Johnson, B. K., Kenya, P. R., Ngira, J. A., Ocheng, D. M., Nandwa, H., Omondi, T. N., Jansen, A. J., Ngare, W., Kaviti, J. N., Gatei, D., and Siongok, T. A., Outbreak of acute hepatitis caused by aflatoxin poisoning in Kenya, *Lancet*, June 12, 1346, 1982.

O'Brien, K., Moss, E., Judah, D., and Neal, G., Metabolic basis of the species difference to aflatoxin B_1 induced hepatotoxicity, *Biochem. Biophys. Res. Comm.*, 114, 813, 1983.

Oda, T., Tsuda, H., Scarpa, A., Sakamoto, M., and Hirohashi, S., Mutation pattern of the p53 gene as a diagnostic marker for multiple hepatocellular carcinoma, *Cancer Res.*, 52, 3674, 1992.

Peers, F. G. and Linsell, C. A., Dietary aflatoxins and liver cancer — a population based study in Kenya, *Brit. J. Cancer*, 27, 473, 1973.

Ramsdell, H. S. and Eaton, D. L., Species susceptibility to aflatoxin B_1 carcinogenesis: Comparative kinetics of microsomal biotransformation, *Cancer Res.*, 50, 615, 1990.

Raney, K. D., Coles, B., Guengerich, F. P., and Harris, T. M., The endo 8, 9-epoxide of aflatoxin B_1: a new metabolite, *Chem. Res. Toxicol.*, 5, 333, 1992.

Reed, D. J. and Ellis, W. W., Effects of *in vivo* inactivation of γ-glutamyl transpeptidase on biliary excretion of glutathione and glutathione conjugates, *Pharmacologist*, 23, 167, 1981.

Ross, R. K., Yuan, J.-M., Yu, M. C., Wogan, G. N., Qian, G.-S., Tu, J.-T., Groopman, J. D., Gao, Y.-T., and Henderson, B. E., Urinary aflatoxin biomarkers and risk of hepatocellular carcinoma, *Lancet*, 339, 943, 1992.

Sabbioni, G., Ambs, S., Wogan, G. N., and Groopman, J. D., The aflatoxin-lysine adduct quantified by high-performance liquid chromatography from human serum albumin samples, *Carcinogenesis*, 11, 2063, 1990.

Sabbioni, G., Skipper, P. L., Büchi, G., and Tannenbaum, S. R., Isolation and characterisation of the major serum albumin adduct formed by aflatoxin B_1 *in vivo* in rats, *Carcinogenesis*, 8, 819, 1987.

Sargeant, K., Sheridan, A., O'Kelly, J., and Caraghan, R. B. A., Toxicity associated with certain samples of ground nut meal, *Nature (London)*, 192, 1095, 1961.

Shimada, T. and Guengerich, F. P., Evidence for cytochrome $P450_{nf}$, the nifedipine oxidase being the principal enzyme involved in the bioactivation of aflatoxins in human liver, *Proc. Natl. Acad. Sci. U.S.A.*, 86, 462, 1989.

Sinha, S., Webber, C., Marshall, C. J., Knowles, M. A., Proctor, A., Barrass, N., and Neal, G. E., Activation of ras oncogene in aflatoxin-induced rat liver carcinogenesis, *Proc. Natl. Acad. Sci. U.S.A.*, 85, 3673, 1988.

Solt, D. and Farber, E., New principle for the analysis of chemical carcinogens, *Nature*, 263, 701, 1976.

Soman, N. R. and Wogan, G. N., Activation of the c-Ki-ras oncogene in aflatoxin B_1-induced hepatocellular carcinoma and adenoma in the rat: Detection by denaturing gradient gel electrophoresis, *Proc. Natl. Acad. Sci. U.S.A.*, 90, 2045, 1993.

Tandon, H. D. and Tandon, B. N., Pathology of liver in an outbreak of aflatoxicosis in man with a report on the follow up, in *Mycotoxins and Phycotoxins '88, A collection of Invited Papers Presented at the Seventh International IUPAC Symposium on Mycotoxins and Phycotoxins, Tokyo, Japan, 16–19 August 1988,* Natori, S., Hashimoto, K., and Ueno, Y., Eds., Bioactive Molecules, Vol. 10, Elsevier, Amsterdam, 1989, 99.

Wild, C. P., Pionneau, F. A., Montesano, R., Mutiro, C. F., and Chetsanga, C. J., Aflatoxin detected in human breast milk by immunoassay, *Int. J. Cancer*, 40, 328, 1987.

Wild, C. P., Jiang, Y.-Z., Sabbioni, G., Chapot, B., and Montesano, R., Evaluation of methods for quantitation of aflatoxin-albumin adducts and their application to human exposure assessment, *Cancer Res.*, 50, 245, 1990.

Chapter 5

EXPERIMENTAL STUDIES WITH FIBERS

J. M. G. Davis and R. C. Brown

CONTENTS

I. Introduction 85

II. Early Studies with Asbestos 85

III. Experiments with Non-asbestos Natural Fibers 88

IV. Experiments with Man-made Mineral Fibers 89

V. The Mechanisms Operating during Fiber Pathogenesis 93

VI. Conclusions 95

References 95

I. INTRODUCTION

Man's awareness of the possible hazard of exposure to high levels of respirable mineral fibers began with asbestos. This group of materials began to be widely used at the end of the 19th century when the consideration of workplace hygiene was almost nonexistent. Consequently exposure was often to massive levels of airborne fibers. Such exposures caused serious, and often fatal, pulmonary interstitial fibrosis with the first formal report of a case being recorded by Murray in 1907 (Murray, 1907). Cooke (1924) first used the term asbestosis to describe this condition and cases were then reported from the asbestos industry in all areas of the world.

II. EARLY STUDIES WITH ASBESTOS

The seriousness of this situation led to the first industrial sponsored experimental studies, which commenced in the 1930s and were undertaken using chrysotile asbestos by Gardner's laboratory at Saranac Lake, New York, in the United States. At this time experimental animal inhalation studies were novel and by modern standards Gardner's study suffered from many deficiencies. However, he did develop techniques for the generation of dense clouds of

0-8493-9229-2/95/$0.00+$.50
© 1995 by CRC Press Inc.

respirable asbestos and used these to treat rats, guinea pigs, mice, rabbits, and cats. He was successful in producing asbestosis in all these species, although there was no real calibration of exposure and certainly no detailed fiber counts. Indeed, since electron microscopy was not available, accurate counts would not have been possible with chrysotile asbestos. Gardner's studies, which were published after his death (Vorwald et al., 1951), did achieve one notable success. He was able to report that finely ground chrysotile, consisting mainly of short fibers, was less able to produce asbestosis than longer chrysotile fibers — this dependence on fiber morphology and size being perhaps *the* essential fact of fiber toxicology.

At the time Gardner's studies were undertaken, it was not realized that asbestos fibers could be carcinogenic as well as fibrogenic, although a few case reports of asbestos workers with both asbestosis and lung cancer had appeared in the literature (Lynch and Smith, 1935). Rats are now recognized as the species most susceptible to fiber lung carcinogenesis, but in Gardner's experiments, they all died relatively early due to pulmonary infection. This led to Gardner's failure to detect any carcinogenic effect of fiber exposure and emphasizes the advantage of modern SPF animals. In his experiments Gardner did produce an apparently raised incidence of tumors in some of the mice he used, but, unfortunately, adequate control animals were not available. It was not until 1957 when Lynch et al. (1957) reported a more detailed inhalation study, again using mice, that this observation was followed up. Evidence of a raised incidence of lung cancer in asbestos workers had by this time become much more definite, being confirmed by Doll (1955). However, Lynch failed to demonstrate a clear carcinogenic effect of treatment with chrysotile; and this could be due to the use of mice, a species now regarded as unsatisfactory for studies examining the carcinogenesis of airborne particles. Holt et al. (1964, 1965) undertook asbestos inhalation studies with rats and guinea pigs. Both species developed asbestosis, but no pulmonary tumors were found. Guinea pigs are also now regarded as unsuitable for carcinogenicity studies, and once again the rats used did not survive long enough for detection of a carcinogenic response.

The first inhalation study which did clearly demonstrate a carcinogenic effect was reported by Gross et al. (1967) who had treated rats with chrysotile asbestos. These animals developed pulmonary fibrosis and both benign and malignant pulmonary tumors. Similar studies were published by Reeves et al. (1971, 1974) who confirmed the suitability of the rat for studies of fiber carcinogenesis. Reeves treated rats, rabbits, guinea pigs, gerbils, and mice with chrysotile, crocidolite, and amosite. While all species tested developed asbestosis, pulmonary tumors were found only in rats.

Wagner, having demonstrated that exposure to crocidolite fibers could cause pleural mesotheliomas in asbestos miners (Wagner et al., 1960), rapidly produced these tumors experimentally by injecting asbestos fibers directly into the pleural cavity of rats (Wagner, 1962). Later Wagner and co-workers carried out a long series of inhalation studies using rats exposed to many types of fiber. The first large study from this group (Wagner et al., 1974) used four types of asbestos (chrysotile, amosite, crocidolite, and anthophyllite) administered in a wide range of doses. These materials were all reference samples produced by the Medical Research Councils of Britain and South Africa whose preparation was financed by the Union International Contre le Cancer (UICC) (Timbrell and Rendall, 1972). Wagner had recognized that dose in an inhalation experiment is a product of exposure concentration and time, and so all the dusts were administered at a constant level of 10 mg/m^3 of air for varying periods up to 2 years. Pulmonary fibrosis and tumors developed with all four dusts, and the level of effect was clearly shown to be related to the cumulative dose of fiber received. A few mesotheliomas developed following treatment with each type of asbestos. This pattern of response has now been confirmed many times. With rats, fiber exposure by inhalation rapidly leads to pulmonary fibrosis and at high dose a large proportion of the animals can develop pulmonary tumors, mesotheliomas do develop but with most types of fiber these are relatively rare even at high doses.

Davis et al. (1978) reported inhalation studies in which rats were exposed to UICC amosite, crocidolite, and Rhodesian chrysotile administered at equal number of fibers >5 μm long (as determined by optical phase contrast microscopy) and also equal mass concentrations. This study demonstrated that at both equal mass and equal fiber number, chrysotile asbestos produced the most fibrosis and tumors. The authors considered that these results were due to the greater number of very long fibers (those more than 10 μm) in the chrysotile dust cloud. The difference between exposure to the two materials would have been more obvious if the size of fiber in the dust clouds had been measured by transmission electron microscopy since more chrysotile fibers would have escaped detection by optical microscopy.

These studies by both Wagner's and Davis's groups demonstrated one major difference between chrysotile and amphibole asbestos. Chrysotile was found to accumulate much less rapidly in the rat's lungs and, after the inhalation period, chrysotile cleared from the lungs much faster. This difference has been confirmed in studies of fiber levels in human autopsy specimens (Gylseth et al., 1983). A further difference between chrysotile and the amphiboles was demonstrated later when found that chrysotile more readily separated into the individual crystal fibrils, particularly in "lung tissue fluids," thus causing at least an initial increase in the number of pathogenic fiber units (Davis et al., 1986a; Bellman et al., 1987).

Gardner's original observation that long asbestos fibers were more carcinogenic than short fibers had been extended by data obtained from studies in which fibers were injected or implanted into either the pleural or peritoneal cavities to show that long thin fibers were also the most carcinogenic (Pott and Friedrichs, 1972; Pott, 1978; Stanton and Wrench, 1972; Stanton et al., 1977). It could not be confirmed that this relationship between fiber length and pathogenic potency was true for the inhalation route until sufficient quantities of suitable dusts became available. Very short fiber samples of both amosite and chrysotile were eventually produced and compared with samples containing high proportions of very long fibers (Davis et al., 1986b; Davis and Jones, 1988). A short fiber preparation of amosite containing virtually no fibers >5 µm in length produced neither fibrosis nor pulmonary tumors, while long-fiber amosite was extremely fibrogenic and carcinogenic. The short-fiber chrysotile preparation available was less satisfactory, having a significant number of fibers >5 µm in length, but fibrosis and tumor production were very much lower than with the long-fiber chrysotile preparation. By using data from two studies, Davis (1989) pointed out that following inhalation, long-fiber amosite was extremely pathogenic, while both the UICC preparation and a very short-fiber preparation produced very little disease. Following intraperitoneal injection, however, both the long-fiber preparation and the UICC specimen produced mesotheliomas in almost 100% of animals, while the short-fiber sample had almost no effect. Since the major difference between the long-fiber amosite and the UICC sample was the presence of numerous very long fibers (>20 µm in length) in the former, it was suggested that this length of fiber is necessary for the production of pulmonary fibrosis and tumors following inhalation. As previously reported (Pott and Friedrichs, 1972; Stanton and Wrench, 1972) fibers >8 µm in length appeared to produce maximum numbers of mesotheliomas following injection. It remains to be determined whether or not mesotheliomas can be produced by the inhalation of this relatively short-fiber material.

III. EXPERIMENTS WITH NON-ASBESTOS NATURAL FIBERS

Several inhalation studies have been undertaken with naturally occurring mineral fibers other than the main commercial asbestos varieties. Davis et al. (1985) examined brucite and tremolite, two materials regularly found in small quantities contaminating chrysotile ore bodies and, on some occasions, as pure substances. Unfortunately the brucite specimen obtained as "pure" proved to contain some chrysotile fibers, so the pathogenic potential of brucite itself could not be determined. The Korean tremolite sample obtained was however very pure and proved to be extremely fibrogenic and carcinogenic. This finding is important since it fits in with some human observations. While tremolite contamination of chrysotile ore is usually extremely small, this material, which is an amphibole, is extremely durable in lung tissue, and in chrysotile miners examined at autopsy the retained tremolite often comprises the bulk of the fiber present (Sebastien et al., 1989). Because of the known potency of amphiboles in mesothelioma production, it has been suggested that this retained tremolite is responsible for the few mesotheliomas reported from chrysotile miners rather than the chrysotile itself (Pooley, 1976; Churg, 1988). Perhaps the most dramatic experimental inhalation study with fibers has been undertaken with fibers of erionite (Wagner et al., 1985). Baris et al. (1979) had reported that an extremely high incidence of mesotheliomas was found in some isolated Turkish villages contaminated with this material, and in Wagner's experiment all of the rats in the exposed group developed mesothelioma. The reason for the carcinogenic potency of erionite remains uncertain. Certainly fiber numbers and dimensions do not explain it. However, even with erionite, a reduction of fiber length below a critical level eliminates the carcinogenic potential (Wagner, 1990).

Several explanations have been put forward to explain this extreme, but still size dependent, activity of erionite. Poole et al. (1983) reported that, unlike asbestos, erionite was genotoxic in several *in vitro* systems. It is also possible that erionite has the ability to penetrate the rat pleural surface, which asbestos may not do without the help of other biologically active dusts (Davis et al., 1991).

IV. EXPERIMENTS WITH MAN-MADE MINERAL FIBERS

While asbestos and erionite were suspected of being harmful to humans before experimental studies commenced, the situation was different with man-made fibers. Here, no human hazard had been detected when both work in the United States (Stanton and Wrench, 1972) and Germany (Pott and Friedrichs, 1972) demonstrated that fine man-made fibers, particularly those made from glass, could cause mesotheliomas following injection or implantation in rats. Fiber dimension seemed to be the only important factor. The possibility that man-made fibers could perhaps be harmful resulted in a series of inhalation studies primarily with glass fiber. Early work on man-made fibers was flawed by the use of unsuitable dust clouds. Inhalation studies by Schepers et al. (Schepers and Delahant, 1955; Schepers, 1959), Gross et al. (1970) and Lee et al. (1979) using rats and other rodents failed to demonstrate a pathogenic potential with "fibrous" glass, but their dust clouds were composed mainly of either particulate fragments or fibers too thick for rats to respire.

Both Wagner et al. (1984) and McConnell et al. (1984) avoided these difficulties by using an extremely fine glass micro fiber with fiber dimensions similar to asbestos. Dust clouds generated from this material contained very high numbers of long thin fibers, but even so no fibrosis or tumor production occurred in exposed rats. The same micro fiber was, however, extremely carcinogenic when injected into the pleural or peritoneal cavities of rats. This was the first demonstration for the man-made fibers of a real, and potentially confusing, difference between results obtained by these routes of administration, a problem still causing difficulties today.

It is probable that part at least of the explanation for these discrepancies may be found in the "durability" of these fibers in the lung and pleural cavity. The level of sufficiently long fibers has to reach some critical level in the lungs before disease results, and this level has to be maintained for a significant proportion of the life-span of the exposed species. The existence of such a threshold is likely for fibrosis, but is more difficult to justify for the malignancies. However, for lung cancer after human asbestos exposure it is probable that the fibers act at a late stage in carcinogenesis (Peto, 1983). For mesothelioma, fibers may act at an early stage and be complete carcinogens or initiators. Fibers deposited in lung tissue are subject to macrophage clearance, but also are readily subjected to chemical dissolution which may be complete or may result in fiber weakening and fracture to shorter and nonpathogenic lengths. Clearly the injection of fibers directly into the body cavities requires no time for build up of a critical dose, and the fibers are deposited in clumps much less susceptible to dissolution and related processes; this may produce tumors where none would occur following inhalation.

The problem of producing suitable dust clouds of man-made fibers for test purposes remains. Many are relatively thick, but bulk samples do contain a small proportion of respirable fibers. The hazardous nature of these fine fibers still needs to be examined, although in use too few of them would become airborne to pose any risk to man. The problem of obtaining a dust cloud with sufficiently high numbers of long thin fibers within an acceptable airborne mass of dust is acute and exacerbated by the fact that only thinner fibers (for a specific gravity of about 2.7, those <1 μm in diameter) can penetrate deeply into the pulmonary system of the rat than is true for humans (<3 μm diameter). A combination of these factors has resulted in several studies whose results appear clear cut, but the implications of which for humans are uncertain.

Le Bouffant et al. (1987), Muhle et al. (1987) and Smith et al. (1987) reported inhalation studies in rats with glass, rockwool, and ceramic fibers. None of these dusts showed evidence of pathogenic effects in rat lungs. However, numbers of respirable fibers were low and sometimes below 100 fibers/ml (>5 μm in length by phase contrast optical microscopy). Fiber diameters in the dust clouds were not reported. The studies of Wagner et al. (1984) and McConnell et al. (1984) did not closely examine the durability of glass micro fiber in lung tissue at the same time that the pathogenic potential of the dust was determined. The importance of this information was recognized by Le Bouffant et al. (1987) and Bellman et al. (1987). These studies found wide differences in the survival of fibers in the pulmonary parenchyma, and many glass preparations disappeared very quickly, while some other materials such as ceramic fibers were more durable. The recognition that "durability" is one of the essential factors in fiber pathogenicity resulted in a conference devoted to this subject in Lyon (Bignon and Sarracci, 1992).

When considering fiber durability and pathogenicity, it is important to allow for differences in life-span between humans and experimental animals. Tumor production, particularly, requires a significant proportion of total life-span. Thus fibers that can survive in lung tissue for 1–2 years may be highly pathogenic in rats, but unable to accumulate fast enough in human lungs to cause disease unless dust levels are exceptionally high. Chrysotile asbestos appears to be a good example of this. It is known to be much less pathogenic to humans than the amphiboles, yet in rat studies it has proved to be equally fibrogenic and carcinogenic as any other asbestos variety.

If fibers are sufficiently durable, it seems that even man-made fibers can produce some disease in rodents if the exposure is sufficient. Davis et al. (1984) treated rats with a dust cloud of refractory ceramic fibers (RCF) of only 100 fibers/ml >5 μm in length, many of which were above 1 μm in diameter. Some indication of both fibrosis and the production of pulmonary tumors was found. In the experiment with ceramic fibers carried out by Smith et al. (1987), who exposed both Osborne-Mendel rats and Syrian Golden hamsters to 200 f/ml for 6 hours/day, 5 days/week for 2 years, no excess of tumors was observed in the rats and little, if any, fibrosis. In the hamsters, there was no fibrosis but one animal developed a pleural mesothelioma.

The doubts about the significance of the results obtained in these two investigations led the RCF industry to sponsor the largest single series of inhalation experiments on fibers yet undertaken, those carried out at the Research and Consulting Company in Geneva (RCC). Previous studies used only readily available industrial ceramic fiber materials milled or ground to reduce overall fiber length. The aerosols produced contained a significant quantity of material that was not respirable by the rodent and fragments of the glassy balls or "shot" that makes up about 50% by weight of these materials. The lung burden of fibers and the sizes of fibers recovered from the exposed animals was not reported. In contrast, the RCFs used in the RCC experiments were treated to remove the shot and to select the finer, rodent-respirable fibers. The size selection was carried out using a proprietary method for which no details are available, but nearly one ton of each commercial product had to be processed to provide the material used in the experiments. Three types of commercial material were used, a kaolin-based fiber, a "high purity fiber" (made from alumina and silica), and a zirconia fiber, containing 15% zirconia. The isolated kaolin fiber was also heated to simulate a devitrified, "after-use" fiber. Animals were exposed in a system that maximized lung burden in relation to the exposure (Bernstein and Drew, 1980) and presumably this also maximized the potential for lung damage.

Following histopathological examination of the lungs after 1 month's exposure of Fischer 344 rats at 20, 40, and 60 mg/m³ of the zirconia fiber, the maximum tolerated dose (MTD) was estimated (Glass et al., 1993) to be 30 mg/m³ for all the test fibers in both the rat and hamster. Both of these species were therefore exposed to this dose (approximately 200 fibers/cc), for 6 hours/day, 5 days/week. Rats were exposed to all four fibers for 24 months, and hamsters to only the kaolin fiber for 18 months. A sample of chrysotile asbestos was included in the study as a presumptive "positive control" at an exposure of 10 mg/m^3. Rats thus exposed developed a lung tumor incidence of 18.5%. Exposure to the MTD of the various chemically different, "size-selected" RCFs caused a small but statistically significant increase in lung tumors.

Thirty-eight percent of the hamsters exposed to the kaolin RCF were found to have mesothelial tumors by microscopic examination at post-mortem. These tumors did not affect survival. No mesothelial tumors were seen in the animals exposed to the chrysotile asbestos. Exposure of hamsters to RCF or chrysotile did not produce lung tumors (Hesterberg et al., 1992; Bunn et al., 1993). Given the apparent sensitivity of hamsters to the pleural effects of sufficiently long ceramic fibers it is regrettable that only the kaolin RCF has been examined at one dose in this species. Other fibers of sufficient length should also be tested in this species as a means to understanding the nature of this, so far, unique response (Bunn et al., 1993).

The study at RCC was extended by exposing three additional groups of rats to different levels of the kaolin RCF (3, 9, and 16 mg/m^3, corresponding to approximately 25, 75, and 115 f/cc). Animals exposed at 16 mg/m^3 developed pleural and parenchymal fibrosis; at 9 mg/m^3 there was mild parenchymal fibrosis, while at the lowest dose there were no irreversible effects. There was no excess of lung tumors at any of these doses, while one rat exposed to 9 mg/m^3 developed a microscopic mesothelioma.

Thus for inhalation exposure, there is evidence of a no-effect level for both fibrosis and tumors, and this is consistent with the possibility that the fiber exposures in these experiments produced inflammatory responses which were associated with epithelial hypertrophy and/or hyperplasia and squamous metaplasia. It could therefore be argued that the lung tumors at the MTD were produced by an epigenetic mechanism mediated by cytotoxicity followed by a consequent regenerative cell proliferation.

The data from the RCC studies of RCF have been used in the model of pulmonary deposition, retention and clearance developed by Yu et al. (1990a, b, 1991, in press a, b) It appears that the estimated MTD was too high, and pulmonary "overload" may have occurred. This could have had a profound impact on the responses of the exposed animals. Results from this model may also provide the basis to explain the observed differences between rats and hamsters and provide extremely useful insights for future quantitative risk assessments of fibers. The need to avoid overload (empirically defined as a particulate burden which exceeds the clearance mechanisms of the lung) in inhalation experiments complicates the interpretation of these experiments since overload particulate-induced lung tumors appear to have no correlate in exposed humans (Hext, 1994; Morrow, 1992; Lewis et al., 1989).

In subsequent experiments at RCC multiple doses of two samples of glass fiber and two samples of mineral wool (one rockwool and one slag wool) have been studied. The only exposure in this series producing significant pathology was fibrosis at the highest dose of rockwool. Samples of glass wool will be examined in hamsters in a future experiment.

The difference between the results obtained with ceramic and other man-made minerals could be due to differences in the size distribution of the fibers, their surface chemistry, or durability. Overall, the retention of all fibers (longer than 5 µm) was similar with all the materials, suggesting that durability is not an issue. However, the retention of very long fibers (more than 20 µm) was highest in the ceramic and rockwool fibers (Brown et al., in press). Morgan et al. (1982) demonstrated that long glass fibers disappear more rapidly from the lungs than do shorter fibers which are contained inside macrophages in an acid environment and thus protected from alkaline dissolution. Ceramic and rockwool fibers show a quite different pattern of pH-dependent solubility, and long ceramic fibers therefore remain in the lung longer than similar glass fibers — though still less than asbestos. Additionally the ceramic fiber contained more nonfibrous respirable particulate, and this could have altered the clearance of the fiber and thus contributed to the production of "overload."

It seems that despite the careful control of the conditions used in the RCC experiments there remain sufficient uncertainties to preclude easy distinction between different man-made fibers. However, it seems clear that only very high doses of even the most durable fiber can produce disease, at least in the rat. A suitable design for animal inhalation studies aimed at predicting human hazard from fibers could include the following: dust clouds should contain at least 100 fibers/ml with a length >20 µm and a diameter of <1 µm. The exposure should not produce a marked reduction in the operation of the normal lung clearance pathways. In rats, inhalation must be continued for at least 2 years, and animals must be allowed to survive for their full life-span before pathological examination. These requirements are quite different from existing international guidelines for general inhalation toxicology, and there is therefore a need for fiber-specific guidelines to be adopted, including appropriate measurements of dose of sized fibers to target tissue.

V. THE MECHANISMS OPERATING DURING FIBER PATHOGENESIS

There are a number of hypotheses seeking to explain the pathogenicity of mineral fibers, but with three disease endpoints to explain — lung cancer, mesothelioma, and fibrosis — there have been attempts to invoke both separate and unifying mechanisms. As described above, the critical feature of all fiber-related disease is the size dependency of the response. This size dependency could be due to retention or localization of the active fibers at the sensitive target sites. However many *in vitro* effects of fibers show a similar size dependency to that seen in the whole animal studies. Thus the effect of size must lie, at least in part, at the cellular level (see review by Petruska et al., 1991). Thus all potential mechanisms must be examined for their size dependency. For many years, the role of oxygen free radicals have been invoked to explain the pathogenicity of fibers with the source of the active species coming either from the oxidant burst accompanying the phagocytosis or attempted phagocytosis of the fiber or from the catalytic activity of the fiber itself. The content of iron or of some other transition elements is of particular interest, since the most active forms of asbestos — the amphiboles — contain substantial quantities of iron. Crocidolite, or blue asbestos, gets its name from the more than 20% of iron contained in its crystalline structure. However, fibers with less iron are equally or more active in both humans and animals, but this does not completely rule out a role for iron, as one of the effects of residence in the body on fibers is their coating with iron-rich proteinaceous material to form ferruginous or asbestos bodies. Clearly this iron coating could itself confer catalytic properties on the fiber surface, particularly at an early stage of coating. Since only surface groups could react with most targets and then only some surface iron would be in the correct reduction or coordination state, the bulk content of iron might not reflect that which is available and active on the surface.

It is also difficult to see how the reaction of surface groups would depend on fiber size unless the role of size is to deliver the catalytically active material to sensitive sites within the cells or tissues. In most systems, the effects of adding an iron-rich mineral fiber are the same as adding other iron compounds, and the magnitude of the effect depends on the availability and reduction state of the iron. However, most forms of iron-containing material do not have the same profound impact on health as does blue asbestos.

If asbestos can catalyze a reaction in a simple cell free system, it may not mean that the same reaction could damage a cellular or tissue target. This consideration makes it mandatory to investigate the production of radical damage *in situ* rather than in artificial systems. Thus mixing DNA in solution with fibers can damage DNA and produce 8-hydroxydeoxyguanosine (Kasai and Nishimura, 1984; Leanderson et al., 1989), but despite every effort, we have been unable to detect this adduct being formed in nuclear DNA in asbestos-treated intact cells.

While the involvement of free radicals is a popular hypothesis to explain all the activities of fibers, there remain serious gaps in the data available before this can be adopted.

Other theories of fiber pathogenesis seek to be less universal. For example, it has been suggested that fibrosis may be due to the release of cytokines from exposed cells — especially macrophages — or due to cytotoxicity followed by wound healing. The compensatory hyperplasia during wound healing could also explain the sensitivity of the damaged lung to other carcinogens as is seen in the synergism between cigarette smoking and asbestos exposure in humans. The cytotoxicity of fibers has been demonstrated to be related to fiber size in many different *in vitro* systems (Donaldson, 1993) and such toxicity is presumably due to physical interference of the phagocytosed, or partially phagocytosed, fibers with the cellular architecture.

Perhaps the most convincing explanation for the genotoxicity of mineral fibers is that an interference with spindle fiber formation could result in aneuploidy from maldistribution of the chromosomes at cell division (Lechner et al., 1986). It is also possible that physical damage to chromosomes could occur at this time.

VI. CONCLUSIONS

The few fibers known to cause disease in man have been extensively studied in animals, and there is good agreement between pathological responses following natural or experimental exposures. However, there remain many questions to be answered about the effects of mineral fibers and the mechanisms responsible for these effects. Fibers of all types remain essential to the efficient operation of industry and even our homes, and perhaps the most serious consequence of this ignorance is that doubt is always being expressed about the safety of these fibrous materials. Assays for predictive toxicology in animals may well exaggerate their effects, as the outcomes from exposure seem to depend on the time fibers persist as a fraction of the life-span of the animal at risk. However, the physical mismatch in the respiratory tracts of humans and laboratory rodents means that a few materials cannot be tested satisfactorily by inhalation in experimental systems. There is, however, a consensus that inhalation and not injection or implantation is the only realistic route for such testing (McClellan et al., 1992), albeit with suitable correction for the differences in respiratory tract properties.

REFERENCES

Baris, Y. I., Artvinli, M., and Sahin, A. A., Environmental mesotheliomata in Turkey, *Ann. N. Y. Acad. Sci.,* 330, 423, 1979.

Bellman, B., Muhle, H., Pott, F., Konig, H., Kloppel, H., and Spurny, K., Persistance of man-made mineral fibres (MMMF) and asbestos in rat lungs, *Ann. Occup. Hyg.,* 31, 693, 1987.

Bernstein, D. M. and Drew, R., Experimental approaches for exposure to sized glass fibers, *Environ. Health Perspect.,* 34, 47, 1980.

Bignon, J. and Sarracci, R., Biopersistance of respirable synthetic fibres and minerals. Report of a Workshop sponsored by the International Agency for Research on Cancer, Lyon 1992, *Environ. Health Perspect.,* 102, Suppl. 5.

Brown, R. C., Hoskins, J. A., and Glass, L. R., The *in vivo* biological activity of ceramic fibres, *Ann. Occup. Hyg.,* in press.

Bunn, W. B., Bender, J. R., Hesterberg, T. W., Chase, G. R., and Konzen, J. L., Recent studies of man-made vitreous fibres, chronic animal inhalation studies, *J. Occup. Med.,* 35, 101, 1993.

Churg, A., Chrysotile, tremolite and malignant mesothelioma in man, *Chest,* 93, 621, 1988.

Cooke, W. E., Fibrosis of the lungs due to the inhalation of asbestos dust, *Br. Med. J.,* ii, 1, 47, 1924.

Davis, J. M. G., Mineral fibre carcinogenesis: experimental data relating to the importance of fibre type, size, deposition, dissolution and migration, in *Non-occupational Exposure to Mineral Fibres,* Bignon, J., Peto, J., and Saracci, R., Eds., International Agency for Research on Cancer, Lyon, 1989, 33.

Davis, J. M. G., Addison, J., Bolton, R. E., Donaldson, K., and Jones, A. D., Inhalation and injection studies in rats using dust samples from chrysotile asbestos prepared by a wet dispersion process, *Br. J. Exptl. Path.,* 67, 113, 1986a.

Davis, J. M. G., Addison, J., Bolton, R. E., Donaldson, K., Jones, A. D., and Smith, T., The pathogenicity of long versus short fibre samples of amosite asbestos administered to rats by inhalation and intraperitoneal injection, *Br. J. Exptl. Path.,* 67, 415, 1986b.

Davis, J. M. G., Addison, J., Bolton, R. E., Donaldson, K., Jones, A. D., and Wright, A., The pathogenic effects of fibrous ceramic aluminium silicate glass administered to rats by inhalation or peritoneal injection, in *Biological Effects of Man-made Mineral Fibres,* WHO Copenhagen, 1984, 303.

Davis, J. M. G., Beckett, S. T., Bolton, R. E., Collings, P., and Middleton, A. P., Mass and number of fibres in the pathogenesis of asbestos-related lung disease in rats, *Br. J. Cancer,* 37, 673, 1978.

Davis, J. M. G., Bolton, R. E., Donaldson, K., Jones, A. D., and Miller B., Inhalation studies on the effects of tremolite and brucite dust in rats, *Carcinogenesis,* 6, 667, 1985.

Davis, J. M. G. and Jones, A. D., Comparisons of the pathogenicity of long and short fibres of chrysotile asbestos in rats, *Br. J. Exptl. Path.,* 69, 717, 1988.

Davis, J. M. G., Jones, A. D., and Miller, B. G., Experimental studies in rats on the effects of asbestos inhalation coupled with the inhalation of titanium dioxide or quartz, *Int. J. Exptl. Path.,* 72, 501, 1991.

Doll, R., Mortality from lung cancer in asbestos workers, *Br. J. Indust. Med.,* 12, 81, 1955.

Donaldson, K., Miller, B. G., Sara, E., Slight, J., and Brown, R. C., Asbestos fibre length dependent injury to alveolar epithelial cells in vitro: role of a fibronectin binding receptor, *Int. J. Exptl. Path.,* 74, 243, 1993.

Glass, L. R., Mast, R. W., Chevalier, J., Hesterberg, T. H., Anderson, R., McConnell, E. E., and Bernstein, D. M., Sub acute (28 day) nose only inhalation study of size selected refractory ceramic fibre (RCF) in male Fischer 344 rats, *The Toxicologist ,* 13, 34, 1993.

Gross, P., De Treville, R. T. P., Tolker, E. B., Kaschak, B. S., and Babyak, M. A., Experimental asbestosis: the development of lung cancer in rats with pulmonary deposits of chrysotile asbestos dust, *Arch. Environ. Health,* 15, 343, 1967.

Gross, P., Kaschak, M., Tolker, E. B., Babyak, M. A., and De Treville, R. T. P., The pulmonary reaction to high concentrations of fibrous glass dust, *Arch. Environ. Health,* 20, 696, 1970.

Gylseth, B., Mowe, G., and Wannag, A., Fibre type and concentration in the lungs of workers in an asbestos cement factory, *Br. J. Indust. Med.,* 40, 375, 1983.

Hesterberg, T. W., Mast, R., McConnell, E. E., Chevalier, J., Bernstein, D. M., Burn, W. B., and Anderson, R., Chronic inhalation toxicity of refractory ceramic fibers in Syrian hamsters, in *Mechanisms of Fibre Carcinogenesis,* Brown, R. C., Hoskins, J. A., and Johnson, N. F., Eds., Plenum Press, New York, 1992, 519.

Hext, P. M., Current perspectives on particulate induced pulmonary tumours, *Human and Experimental Toxicology,* 13, 700, 1994.

Holt, P. F., Mills, J., and Young, D. K., The early effects of chrysotile asbestos dust on the rat lung, *J. Path. Bact.,* 87, 15, 1964.

Holt, P. F., Mills, J., and Young, D. K., Asbestosis with four types of fibres, *Ann. N.Y. Acad. Sci.,* 132, 87, 1965.

Kasai, H. and Nishimura, S., DNA damage by asbestos in the presence of hydrogen peroxide, *Gann,* 75, 841 1984.

Le Bouffant, L., Daniel, H., Henin, J. P., Martin, J. C., Normand, C., Tichoux, G., and Trolard, F., Experimental study on long-term effects of inhaled MMMF on the lungs of rats, *Ann. Occup. Hyg.,* 31, 765, 1987.

Leanderson, P., Södverkist, O., and Taggeson, C., Hydroxyl radical mediated DNA base modification by man made mineral fibres, *Br. J. Indust. Med.,* 46, 435, 1989.

Lechner, J. F., Tokiwa, T., LaVeck, M., Benedict, W. F., Banks-Schlegel, S., Yeager, H., Banerjee, A., and Harris, C. C., Asbestos associated chromosomal changes in human mesothelial cells, *Proc. Nat. Acad. Sci. U.S.A.,* 82, 3884, 1986.

Lee, K. P., Barras, C. E., Griffith, F. D., and Waritz, R. S., Pulmonary response to glass fiber by inhalation exposure, *Lab. Invest.,* 40, 123, 1979.

Lewis, T. R., Morrow, P. E., McClellan, R. O., Raabe, O. G., Kennedy, G. R., Schwartz, B. A., Coche, T. J., Roycroft, J. H., and Chabra, R. S., Establishing aerosol exposure concentrations for inhalation toxicity studies, *Toxicol. Appl. Pharmacol.,* 99, 377, 1989.

Lynch, K. M., McIver, F. A., and Cain, J. R., Pulmonary tumours in mice exposed to asbestos dust, *AMA Arch. Indust. Health,* 15, 207, 1957.

Lynch, K. M. and Smith, W. A., Pulmonary asbestosis (iii). Carcinoma of the lung in asbestos silicosis, *Am. J. Cancer,* 24, 56, 1935.

McClellan, R. O., Miller, F. J., Hesterberg, T. W., Warheit, D. B., Bunn, W. B., Kane, A., Lippmann, M., Mast, R. W., McConnell, E. E., and Reinhardt, C. F., Approaches to evaluating the toxicity and carcinogenicity of man-made fibres: summary of a workshop held November 11–13, 1991, Durham, North Carolina, *Reg. Toxicol. Pharmacol.,* 16, 321, 1992.

McConnell, E. E., Wagner, J. C., Skidmore, J. W., and Moore, J. A., A comparative study of the fibrogenic and carcinogenic effects of UICC Canadian chrysotile B asbestos and glass microfibre (JM100), in *Biological Effects of Man-made Mineral Fibres,* WHO, Copenhagen, 1984, 234.

Morgan, A., Holmes, A., and Davison, W., Clearance of sized glass fibres from the rat lung and their solubility *in vivo, Ann. Occup. Hyg.,* 25, 317, 1982.

Morrow, P. E., Dust overloading of the lungs: update and appraisal, *Toxicol. Appl. Pharmacol.,* 113, 1, 1992.

Muhle, H., Pott, F., Bellman, B., Takenaka, S., and Ziem, V., Inhalation and injection experiments in rats to test the carcinogenicity of MMMF, *Ann. Occup. Hyg.,* 31, 755, 1987.

Murray, H. M., Report of Departmental Committee on Compensation for Industrial Diseases, Minutes of Evidence, Cd 3946 127-128 HMSO London, 1907.

Peto, J. Dose and time relationships for lung cancer and mesothelioma in relation to smoking and asbestos exposure, in *Proceedings of a Symposium on Asbestos,* Bundesgesundheitsamt, Berlin, 1982, Dietrich Reimer Verlag 1983, 126.

Petruska, J. M., Mossman, B. T., Brown, R. C., and Hoskins, J. A., Mechanism of action of mineral fibres as studied by *in vitro* methods, in *Mineral Fibres and Health,* Liddell, D. and Miller, K., Eds., CRC Press, Boca Raton, FL, 1991, 303.

Poole, A., Brown, R. C., Turver, C. J., Skidmore, J. W., and Griffiths, D. M., *In vitro* genotoxic activities of fibrous erionite, *Br. J. Cancer,* 47, 697, 1983.

Pooley, F. D., An examination of the fibrous mineral content of asbestos lung tissue from the Canadian chrysotile mining industry, *Environ. Res.,* 12, 281, 1976.

Pott, F., Some aspects of the dosimetry of the carcinogenic potency of asbestos and other fibrous dusts, *Staub-Reinholt Luft,* 38, 486, 1978.

Pott, F. and Friedrichs, K. H., Tumours in rats after intraperitoneal injection of asbestos, *Naturwissenschaft.,* 59, 318, 1972.

Reeves, A. L., Puro, H. E., and Smith, R. G., Inhalation carcinogenesis from various forms of asbestos, *Environ. Res.,* 8, 178, 1974.

Reeves, A. L., Puro, H. E., Smith, R. G., and Vorwald, A. J., Experimental asbestos carcinogenesis, *Environ. Res.,* 4, 496, 1971.

Schepers, G. W. H., Pulmonary histologic reactions to inhaled fiberglass-plastic dust, *Am. J. Path.,* 35, 1169, 1959.

Schepers, G. W. H. and Delahant, A. B., An experimental study of the effects of glass wool on animal lungs, *AMA Arch. Indust. Health,* 12, 276, 1955.

Sebastien, P., McDonald, J. C., McDonald, A. D., Case, B., and Harley R., Respiratory cancer in chrysotile textile and mining industries: exposure inferences from lung analysis, *Br. J. Indust. Med.,* 46, 180, 1989.

Smith, D. M., Ortiz, L. W., Archuleta, R. F., and Johnson, N. F., Long-term health effects in hamsters and rats exposed chronically to man-made vitreous fibers, *Ann. Occup. Hyg.,* 31, 731, 1987.

Stanton, M. F., Layard, M., Tegeris, A., Miller, M., and Kent, E., Carcinogenicity of fibrous glass: pleural response in the rat in relation to fibre dimension, *J. Nat. Cancer Inst.,* 58, 587, 1977.

Stanton, M. F. and Wrench, C., Mechanisms of mesothelioma induction with asbestos and fibrous glass, *J. Nat. Cancer Inst.,* 48, 797, 1972.

Timbrell, V. and Rendall, R. E. G., Preparation of the UICC standard reference samples of asbestos, *Powder Technology,* 5, 279, 1972.

Vorwald, A. J., Durkan, T. M., and Pratt P. C., Experimental studies of asbestosis, *AMA Arch. Indust. Hyg. Occup. Med.,* 3, 1, 1951.

Wagner, J. C., Experimental production of mesothelial tumours of the pleura by implantation of dusts in laboratory animals, *Nature (London),* 196, 180, 1962.

Wagner, J. C., Biological effects of short fibres, in *Proceedings of the VIIth International Pneumoconiosis Conference,* U.S. Department of Health and Human Services, DHHS (NIOSH) Publication No. 90-108, Part Ii, 1990.

Wagner, J. C., Berry, G. B., Hill, R. J., Munday, D. E., and Skidmore, J. W., Animal experiments with MMM(V)F. Effects of inhalation and intraperitoneal inoculation in rats, in *Biological Effects of Man-made Mineral Fibres,* WHO, Copenhagen, 1984, 207.

Wagner, J. C., Berry, G., Skidmore, J. W., and Timbrell, V., The effects of the inhalation of asbestos in rats, *Br. J. Cancer,* 29, 252, 1974.

Wagner, J. C., Skidmore, J. W., Hill, R. J., and Griffiths, D. M., Erionite exposure and mesothelioma in rats, *Br. J. Cancer,* 51, 727, 1985.

Wagner, J. C., Sleggs, C. A., and Marchand, P., Diffuse pleural mesotheliomata and asbestos exposure in North Western Cape Province, *Br. J. Indust. Med.,* 17, 260, 1960.

Yu, C. P. and Asgharian, B., A kinetic model of alveolar clearance of amosite asbestos fibers from the rat lung at high lung burdens, *J. Aerosol. Sci.,* 21, 21, 1990b.

Yu, C. P., Asgharian, B., and Abraham, J. L., Mathematical modelling of alveolar clearance of chrysotile asbestos fibers from rat lungs, *J. Aerosol. Sci.,* 21, 587, 1990a.

Yu, C. P., Asgharian, B., and Pinkerton, K. E., Intrapulmonary deposition and retention modelling of chrysotile fibers in rats, *J. Aerosol. Sci.,* 22, 757, 1991.

Yu, C. P., Zhang, L., Oberdorster, G., Mast, R. W., Glass, L. R., and Utell, M. J., Clearance of refractory ceramic fibres (RCF) from the rat lung: development of a model, *Environ. Res.,* in press.

Yu, C. P., Zhang, L., Oberdorster, G., Mast, R. W., Glass, L. R., and Utell, M. J., Deposition modelling of refractory ceramic fibres in the rat lung, *J. Aerosol. Sci.,* in press.

Part II
Biomonitoring and Molecular
Interactions in Carcinogenesis

Chapter 6

BIOMONITORING AND MOLECULAR DOSIMETRY OF GENOTOXIC CARCINOGENS

Peter B. Farmer

CONTENTS

I. Introduction 101

II. DNA-Carcinogen Adducts 103
    A. ^{32}P-Postlabeling 105
    B. Immunoassay 108
    C. Fluorescence Assays 109
    D. Mass Spectrometry 109
    E. Accelerator Mass Spectrometry 109
    F. Urinalysis 110

III. Protein-Carcinogen Adducts 111
    A. Albumin-Carcinogen Adducts 112
    B. Hemoglobin N-Terminal Valine Adducts 113
    C. Hemoglobin Cysteine Adducts 113
    D. Hemoglobin Histidine Adducts 114
    E. Hemoglobin Carboxylic Acid Adducts 114

IV. Dose-Response Relationship for Hemoglobin
    Adducts and DNA Adducts 115

V. Background Levels of Adducts 116

VI. Conclusion 116

References 117

I. INTRODUCTION

Epidemiology has been successful in identifying over 50 compounds of proven human carcinogenicity (Tomatis et al., 1989). The chemical exposures which led to the identification of these carcinogens were generally high and of an occupational or medicinal nature. The more commonly occurring situation in which man is exposed to complex mixtures of carcinogens at low dose (such

0-8493-9229-2/95/$0.00+$.50

as is the case for diet or urban pollution) is clearly much more complex, and it is unlikely that epidemiology will have an equivalent impact in identifying the carcinogenic contribution of individual components in such mixtures. As the current belief is that up to 35% of human cancers are caused by factors in the diet or general environment (Doll and Peto, 1981), there is first a need to identify the carcinogens in these sources, and second, to categorize them according to their relative carcinogenic risk. One of the essential parameters for the latter is an accurate assessment of exposure; the lack of adequate exposure data has often been a major limiting factor in previous studies of environmental hazards.

Under certain circumstances, reasonably accurate estimations of exposure may be obtained by analysis of the component of interest in the air, food, water, etc., coupled with determinations of the intake by an individual. This is particularly successful for food-derived toxins, e.g., aflatoxin B_1, where meals duplicating those consumed by the individual under study may be prepared and subsequently analyzed to indicate the intake of the toxin (Rees and Tennant, 1993). Airborne pollutants may be similarly assessed, although the estimate of air intake and the potential variability of air concentrations may lead to some uncertainties in the exposure determination.

Although information of value may be obtained from such external exposure measurements, an individual's response will depend upon the distribution of the chemical in the body, and its metabolism and excretion profiles. For these reasons, the determination of the internal dose of the compound or of its metabolites, by measurement of plasma or urine concentrations, holds several advantages. This approach is extensively used for monitoring occupational exposure to potentially carcinogenic compounds, e.g., to styrene by measurement of urinary mandelic acid (Engström et al., 1978), to polycyclic aromatic hydrocarbons by measurement of 1-hydroxypyrene (Jongeneelen et al., 1986; Buchet et al., 1992), and to 4,4′-methylenedianiline (MDA) by measurement of urinary MDA and its N-acetyl metabolite (Cocker et al., 1988).

The data acquired in this way are informative regarding, for example, the circulating concentrations of active metabolites and may be used to detect individuals with genetically determined susceptibility to the carcinogen. As indicated in Figure 6.1, internal dose measurements are an improvement over external dose measurements as an indicator of carcinogenic risk but correspondingly are less useful as an external dose monitor.

A more accurate indication of the biologically effective dose of a carcinogen received by an individual would be obtained by measurement of the extent of interaction of the active chemical (or metabolite) with its target site in DNA. With our present state of knowledge, this is not feasible for two reasons. First, there are only a few chemicals where specific site-selective interactions with DNA have been identified, e.g., activation of the Ha-Ras-1 oncogene at codon 12 by N-methyl-N-nitrosourea (Zarbl et al., 1985). Second, even if the knowledge of the target site was available, the methods available for monitoring

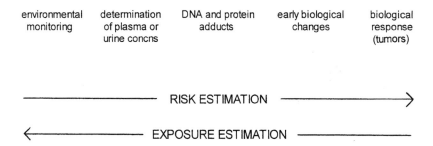

FIGURE 6.1.   Biomarkers of chemical carcinogenesis.

genetic damage at that site are limited, i.e., even if a specific mutation could be detected by, say, a polymerase chain reaction technique, the evidence of the chemical that caused it would not be available.

The methods that most closely approach a determination of biologically effective dose of genotoxic carcinogens depend upon the analysis of adducts (covalently bound products) of the carcinogen with DNA or protein. Although the available analyses may be for a particular chemical structure of adduct, they are not sequence-selective, i.e., they will measure all of the adducts of that structure in the macromolecule. This gives the advantage that sufficient adduct is available for chemical analysis, but clearly, although close, it is not a "target-site" measurement. The outstanding future challenge for those in the carcinogen-biomarker field is to marry the physicochemical analysis of adducts with the mutational analysis so that, following an exposure, both a quantitative estimate of mutagenic risk may be given and the chemical involved in causing the mutation is identified. At present, however, great advances are being made in the development of techniques for measuring adducts in DNA and non-target sites such as protein, and it is on these that this review will largely concentrate.

II. DNA-CARCINOGEN ADDUCTS

The inherent chemistry involved in the formation of DNA-carcinogen adducts is generally alkylation of an electron-rich site within a DNA base or of the phosphate oxygens. The particular site of attack by the carcinogen on the DNA is dependent upon a variety of electronic and steric factors. Within whole DNA there is considerable sequence selectivity for adduct formation. Thus, for example, simple alkylating carcinogens (e.g., methylating agents) prefer to react with guanine residues preceded by another purine (especially guanine) base (Burns et al., 1987; Glickman et al., 1987). GC-rich regions in genes are thus thought to be particular targets for attack by alkylating agents (Mattes et al., 1988; Hartley et al., 1988). The reason for this is that the neighboring purines electronically enhance the nucleophilicity (electron-donating) properties of the central guanine, making it more active for reaction with electrophilic carcinogens. The site within the DNA strand where carcinogens, especially those of larger molecular structure, interact is also governed by the shape and conformation of the carcinogen molecule, as exemplified by the positional and optical isomers of 7,8-dihydroxy-9,10-epoxy-7,8,9,10-tetrahydrobenzo(a)pyrene (Cosman et al., 1993, and references therein).

There is also a multitude of possibilities for the site of adduct formation within each nucleic acid base structure. For guanine, which is commonly the predominant site of attack for chemical carcinogens, there are six feasible sites (N-1, N-3, N-7, O^6, N^2, C-8) where adduct formation may take place. The relative proportion of reaction at each of these sites is governed by the electronic (and steric) nature of the alkylating agent. Low molecular weight alkylating agents which react by a bimolecular nucleophilic substitution mechanism (S_N2), e.g. dimethyl sulphate, form adducts predominantly at N-7 of guanine. Unimolecular reactants, such as the methylating agent derived from N-methyl-N-nitrosourea, are not as dependent upon the nucleophilicity of the atom in the guanine base, and give a higher proportion of O^6 adducts (Lawley, 1980). Alkylating agents formed from metabolism of polycyclic aromatic hydrocarbons, e.g., benzo(a)pyrene, yield particularly N^2 adducts, and aromatic amines C-8 adducts. Adducts at essentially all of the oxygen and nitrogen atoms in the four DNA bases — guanine, adenine, cytosine, and thymine — have been detected following carcinogen exposure, in addition to phosphotriesters resulting from alkylation of the phosphate oxygen.

The stability of the adducts is very variable, indicating that the timing of sample collection is an important parameter in biomonitoring. Some adducts are chemically labile, e.g., N-alkylated purines are susceptible to hydrolysis of the N-9 glycoside bond. Of more importance is the existence of DNA repair mechanisms which enzymatically remove either the carcinogen-modified base from the DNA (e.g., alkylpurine-DNA glycosylases) or the alkyl group from the base, restoring it to its original state (e.g., O^6-alkylguanine-DNA alkyltransferase). The choice of adduct to be used in a biomonitoring procedure will clearly depend in part on its lifetime. For example, N-7-alkylguanines and N-3-alkyladenines are removed from DNA by glycosylases and are excreted in urine within a few days of their formation. They can thus only be used for recent exposure monitoring. Other alkylated purines have longer lifetimes and may be used for longer retrospective monitoring, and although not widely studied, the phosphotriesters appear to have potential for very long-term retrospective exposure monitoring (Den Engelse et al., 1986; Shooter, 1978).

The extent of DNA-carcinogen adduct formation in human samples is normally less than 1 carcinogen-modified base per 10^6 normal bases. Highly sensitive techniques are therefore required for human biomonitoring. As DNA is not generally very accessible from exposed individuals, the amount of adduct available for analysis is very limited. The most common source of DNA for human biomonitoring is white blood cells or peripheral blood lymphocytes. Sample sizes up to approximately 100 μg may be relatively easily obtained in this way. Other sources that have been explored are placenta, buccal smears,

TABLE 6.1
Analytical Methods for DNA-Carcinogen Adducts, in Increasing Order of Sensitivity but Decreasing Order of Ability for Structure Determination

Method	Sensitivity[a] (adducts/normal base)
Gas chromatography-mass spectrometry	$1/10^6$–$1/10^7$
HPLC - electrochemical detection	$3/10^7$
HPLC - synchronous fluorescence spectrometry	$2/10^8$
Immunoassay	$1/10^7$–$1/10^8$
^{32}P-postlabeling	$1/10^{10}$
Accelerator mass spectrometry	$1/10^{14}$

[a] The sensitivity figures are representative of examples given in the text, to which reference should be made for details. Limits of sensitivity vary extensively between different adducts.

excreted urothelial cells from the bladder, oral biopsies, lung tissue, and other autopsy material. As indicated below, urine is also a convenient noninvasive source of repair products derived from carcinogen-modified DNA.

Five main analytical approaches have been employed for DNA adduct detection and quantitation. These are ^{32}P-postlabeling, immunoassays, mass spectrometry, electrochemical detection, and fluorescence (Table 6.1). The most sensitive of these for unlabeled carcinogen adducts is ^{32}P-postlabeling. However, the sacrifice in achieving high sensitivity is a decrease in identifying power for the adducts. The less sensitive techniques, e.g., gas chromatography-mass spectrometry, can also characterize the chemical nature of the adduct and, hence in most cases, the carcinogen which produced it. If the carcinogen is labeled with ^{14}C, accelerator mass spectrometry (which is considerably more sensitive than ^{32}P-postlabeling) may be used to detect the ^{14}C-isotope in the adduct. Adduct formation by metal-containing carcinogens has been demonstrated using atomic absorption spectrometry.

A. ^{32}P-POSTLABELING

The outline of the ^{32}P-postlabeling method, which was developed originally by Randerath et al. (1981), is shown in Figure 6.2. The adducted DNA is enzymatically hydrolyzed to deoxynucleoside 3'-monophosphates using micrococcal nuclease and spleen phosphodiesterase. This mixture (which contains a vast excess of normal deoxynucleotides compared to adducted deoxynucleotides) is then phosphorylated using ^{32}P-ATP of high specific activity and T_4 polynucleotide kinase, yielding a mixture of 3',5'-bisphosphates. In the original method, this mixture was chromatographed using two-dimensional thin layer chromatography, and the resolved adducts were then detected by autoradiography or Cerenkov counting. The sensitivity of the procedure has now been improved by the incorporation of "enrichment" procedures for the adducted deoxynucleotides prior to their chromatographic separation. For example, butanol extraction selectively concentrates adducts, notably those with aromatic amines and polycyclic aromatic hydrocarbons. Alternatively, nuclease P_1 treatment may be used which removes the phosphate from normal 3'-monophosphates, leaving some adducted phosphates intact, e.g., those with polycyclic aromatic hydrocarbons. Consequently, the normals can no longer be postlabeled, resulting in considerably improved sensitivity for the detection of adducts. Other enrichment approaches for adducts are the use of HPLC or immunoaffinity chromatography for their purification prior to labeling.

The ^{32}P-postlabeling assay has a sensitivity of 1 adduct per 10^{10} normal nucleotides and detects nearly all known genotoxic carcinogens (Table 6.2). Its main disadvantage is that adducts in cases of unknown exposures may not be directly chemically identified, although mass spectrometric methods are now under development for this purpose (Kaur et al., 1993).

The applications of the assay have been extensive (see review by Beach and Gupta, 1992). Of particular significance to human biomonitoring was the recent use of postlabeling to detect exposure to aromatic carcinogens in envi-

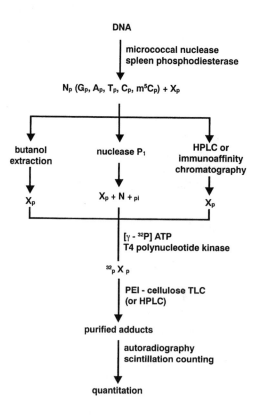

FIGURE 6.2. The ^{32}P-postlabeling assay for DNA-carcinogen adducts.

TABLE 6.2
Determination of DNA-Carcinogen Adducts

Analyte	Analytical method	Source of sample	Carcinogens studied	Recent applications
Adducted 2'-deoxynucleotide	^{32}P-postlabeling	DNA from any available tissue (notably lymphocytes)	Almost all genotoxic carcinogens	Environmental PAHs (Perera et al. 1992) Tamoxifen (White et al. 1992) Styrene oxide (Vodicka et al. 1993)
Alkylpurines	GC-MS immunoassay	Urine	Low molecular weight alkylating agents, aflatoxin B_1	Alkylating agents (Prevost et al. 1993) Aflatoxin B_1 (Ross et al. 1992)
	GC-MS, HPLC-ECD	DNA hydrolysate	Alkylating agents	Ethylene oxide (Walker et al. 1993) Ethylating agents (Van Delft et al. 1993)
Carcinogen moiety	GC-MS, immunoassay, fluorescence	DNA from any available tissue	PAHs	Benzo(a)pyrene (Weston et al. 1989) Benzo(a)pyrene (Weston et al. 1993)
DNA	Accelerator mass spectrometry	DNA from any available tissue (n.b. must be ^{14}C-radiolabeled)	Heterocyclic amines Genotoxic drugs	MeIQ$_x$ (Turteltaub et al. 1990) Tamoxifen (White, Turteltaub et al. unpublished data)
	Immunoassay	DNA from any available tissue (n.b. may require digestion)	Aromatic amines, PAHs, mycotoxins, and many other genotoxins	Ethylating agents (Van Delft et al. 1993) PAH (Kriek et al. 1993) PAH (Mumford et al. 1993)
	Atomic absorption spectrometry	DNA from any available tissue	Metals	Cisplatin (Poirier et al. 1993)

ronmental pollution (Perera et al., 1992). In an exposed Polish population, postlabeling was used to demonstrate an increase in aromatic DNA adducts, notably in winter (with 5.5 ± 4.5 adducts per 10^8 nucleotides) when there is intensive combustion of coal for residential heating.

Postlabeling has also had a major role to play in our recent studies of drug carcinogenicity. Tamoxifen is an antiestrogenic drug used for the treatment of breast cancer, but now proposed for prophylactic treatment of healthy women with an increased risk of this disease. Tamoxifen is a liver carcinogen in the rat, but not in the mouse. Postlabeling has been used (White et al., 1992) to determine DNA adducts produced by tamoxifen in liver DNA of rats (three strains of different susceptibilities), mice, monkeys, and breast cancer patients. The data suggest that humans (and monkeys) exposed to tamoxifen are quantitatively much less susceptible to DNA adduct formation than rats. Postlabeling is now being used in attempts to identify the active metabolite responsible for the DNA damage in the rat. Information obtained from experiments of this type should give clear guidance regarding the risk undertaken by healthy women receiving tamoxifen.

Further applications of ^{32}P-postlabeling for human biomonitoring are summarized in Table 6.2, and reviewed in detail in Beach and Gupta (1992).

B. IMMUNOASSAY

Immunochemical procedures are less sensitive for quantitative detection of DNA-carcinogen adducts than ^{32}P-postlabeling, but hold the advantage of their relative simplicity and lack of expense. As indicated in Table 6.2, antibodies (monoclonal and polyclonal) have been generated against a variety of carcinogen adducts, although some of these (e.g., that against $7\beta,8\alpha$-dihydroxy-9α, 10α-epoxy-7,8,9,10-tetrahydrobenzo(a)pyrene modified DNA) show crossreactivity against structurally-related adducts (Poirier, 1991). This antibody has had extensive use in monitoring occupational and environmental exposures to benzo(a)pyrene (Everson et al., 1986; Harris et al., 1985; Poirier, 1991) and was used, for example, to detect adducts in the environmentally polluted Polish population referred to in the above section (30.4 ± 31.7 adducts per 10^8 nucleotides in winter) (Perera et al., 1992). Other applications of immunochemical biomonitoring (Table 6.2) have been in the measurement of exposure to drugs, e.g., cisplatin, to aflatoxin B_1, to environmental or smoking-related aromatic amines, and to methylating, ethylating, and 2-hydroxyethylating agents. The analytical procedures used have included enzyme linked immunosorbent assay (ELISA), competitive radioimmunoassay (RIA), and ultrasensitive enzymatic radioimmunoassay (USERIA). Another extremely important use of the antibodies lies in their use for production of immunoaffinity resins. These have great potential for the separation of adducts prior to their analysis by other techniques, e.g., ^{32}P-postlabeling (Cooper et al., 1992) and urinalysis (Prevost et al., 1993). For such purposes, it may be advantageous to produce and use antibodies with broad structure specificity (class-specificity), so that a variety of structurally related adducts may be simultaneously isolated.

Further details of the use of immunoassays for the detection of carcinogen-DNA adducts are reviewed by Poirier (1991).

C. FLUORESCENCE ASSAYS

Although fluorescence measurements are limited to selected adducts, they have sufficient sensitivity to measure less than 1 adduct per 10^7 bases. Their application has been largely to exposures to aflatoxin B_1 and polycyclic aromatic hydrocarbons. The sensitivity of the fluorescence measurement may be increased by using synchronous scanning of excitation and emission wavelengths (SFS). A high degree of purification of the sample is required; for example, the quantification of benzo(a)pyrene diol epoxide (BPDE) adducts with placental DNA (Manchester et al., 1988) involved the hydrolysis of the adduct followed by purification of the released benzo(a)pyrene tetrahydrotetrol by immunoaffinity chromatography and then HPLC. Similar purification procedures have recently been used by Weston et al. (1993) in studies of DNA from lung tissues. SFS showed the presence of 1 to 40 BPDE adducts per 10^8 nucleotides. Although most alkylpurines are only weakly fluorescent, there is considerable potential in the use of derivatization procedures to increase their fluorescence and hence lower their detection limits. Thus highly fluorescent products are obtained through reaction of 7-alkylated guanine with phenylmalondialdehyde, and methods for detecting these adducts in DNA are being developed (Sabbioni et al., 1986; Shuker and Durand, 1992).

D. MASS SPECTROMETRY

The highest sensitivity achievable using conventional organic mass spectral techniques is obtained using selected ion recording of the analyte, normally during its chromatography on a GC or HPLC column. Extra selectivity in the analysis may be achieved using tandem mass spectrometry (MS-MS) although this involves a sacrifice in sensitivity. The advantage of the use of mass spectrometry is that it gives chemical structural information on the adduct. After the structure has been identified, quantitative methods for the determination of the adduct may be established using stable isotope-labeled internal standards. The application of mass spectrometry in biomonitoring human DNA-carcinogen adducts has been somewhat limited to date (largely because of the lack of sufficient DNA to analyze) (Table 6.2). However, for example, GC-MS procedures for urinary alkylated purines (see Urinalysis), oxidized DNA bases (Dizdaroglu, 1986), benzo(a)pyrene tetrahydrotetrol (liberated by hydrolysis of BPDE adducts) (Manchester et al., 1988; Weston et al., 1989), and N^2-3-ethenoguanine (an adduct from vinyl chloride exposure) (Fedtke et al., 1990) have all been reported (see review by Fedtke and Swenberg, 1991).

E. ACCELERATOR MASS SPECTROMETRY (AMS)

An accelerator mass spectrometer is an instrument for the highly sensitive detection of isotopes, originally developed for the study of long-lived cosmogenic isotopes, e.g., ^{14}C used for radiocarbon dating. One approach to detect

[14]C in an organic sample involves, firstly, its conversion to carbon. This is inserted into the MS source, where it is ionized to C[-]. These ions are passed into a tandem electrostatic Van de Graaff accelerator, and after acceleration pass through a stripping foil which removes electrons, yielding C[4+], which is then further accelerated. [14]C[4+] then passes through a variety of analysis and focusing sectors prior to its detection. Integration of the beam is achieved by the measurement of the more abundant isotope [13]C. From the ratio of [14]C:[13]C, the fraction of ^{14}C in the original sample may be measured, to extremely high sensitivity.

The application of AMS to biomonitoring has been pioneered by Turteltaub and Felton (c.f. Turteltaub et al., 1990; Felton et al., 1990). Using ^{14}C-labeled carcinogens, the sensitivity of AMS is such that it can detect 1 adduct per 10^{14} nucleotides, considerably more sensitive than ^{32}P-postlabeling (or scintillation counting). In practice, the sensitivity is normally limited by the natural occurrence of ^{14}C at the level of 10^{-12}. Accelerator mass spectrometry may thus be used to measure the relationship between carcinogen dose and DNA adduct levels at much lower doses than had previously been used. For example, the extent of formation of adducts of 2-amino-3,8-dimethylimidazo [4,5-*f*] quinoxaline (MeIQx) with hepatic DNA of mice was found to be linear with respect to dose down to an exposure of 500 ng/kg body weight. The disadvantage of AMS for human biomonitoring is (a) that it cannot detect molecular species, i.e., it gives no information on the structure of the adduct (c.f. ^{32}P-postlabeling) and (b) that ^{14}C labeled material is required. Thus, although a fascinating technique of huge potential for many aspects of biomedical science, it does not seem as yet to be applicable for routine human biomonitoring. There may, however, be interesting experimental possibilities in which very low doses of labeled carcinogens might be administered to volunteers, in order to study the mechanism of adduct formation.

F. URINALYSIS

The presence of modified nucleic acid bases in urine has been known for about 100 years (Krüger and Salomon, 1898). Some of these bases are derived from turnover of tRNA which contains a small proportion of a variety of alkylated bases (Hall, 1971). However, it has also been shown that carcinogen adducts at the N-7 position of guanine and the N-3 position of adenine are unstable within DNA and are removed both by spontaneous chemical hydrolysis of the glycosidic link and by enzymic glycosylases. Thus, for example, the action of human 3-methyladenine glycosylase results in the depurination of both 3-methyladenine and 7-methylguanine, largely within 24 hours of their formation. The fate of such alkylated purines is normally urinary excretion of the intact molecule, although studies by Prevost et al. (1993) using radiolabeled compounds indicate that metabolism may also occur, e.g., 3-benzyl adenine.

The quantitation of urinary alkylated purines is a convenient and noninvasive way to detect recent (0 to 2 days) production of carcinogen adducts with nucleic acids. Urine gives access to evidence of DNA damage in the whole body (unlike, say, the postlabeling assay, which uses an aliquot of DNA from a defined site). The sensitivity requirements for urinalysis of alkylated purines might, therefore, be expected to be less rigorous than for other measures of DNA damage. However, the modified purines are diluted in a vast excess of complex urinary components, and sophisticated clean-up procedures are needed prior to analysis of these purines. HPLC is one possibility (Autrup et al., 1987; Cushnir et al., 1993), but immunoaffinity columns appear to be highly suitable for these purifications (Groopman et al., 1992a, b; Prevost et al., 1993). Antibodies have been prepared against 7-alkylguanines, e.g., the methyl (Shuker, 1988) and the aflatoxin B_1 adduct (Groopman et al., 1984) and 3-alkyladenines, e.g., the methyl (Friesen et al., 1991) and ethyl adduct (Eberle et al., 1990). Immunoaffinity columns capable of class-specific isolation of 3-alkyladenines have now been prepared (Prevost et al., 1993). The most appropriate methods for analysis of the purified alkylpurines appear to be UV, fluorescence, GC-MS, and immunoassay (reviewed in Shuker and Farmer, 1992), and these have been applied to biomonitoring human exposures to many carcinogenic agents, including alkylating drugs, alkylating agents in tobacco smoke, and aflatoxin B_1. For the latter case, a study of a population (18,244) in Shanghai showed that individuals with liver cancer excreted more of the aflatoxin B_1 adduct with guanine N-7 than did controls (RR 4.9, 95% CI = 1.5-16.3) (Ross et al., 1992). The amount of N-3-ethyladenine in urine has recently been shown to be increased by cigarette smoking, indicating the presence of ethylating genotoxins in the smoke (Prevost and Shuker, 1992). Further details of the application of urinalysis of alkylated purines are shown in Table 6.2.

III. PROTEIN-CARCINOGEN ADDUCTS

Adducts of carcinogens with proteins are not normally associated with toxicity or carcinogenicity (the exception being some vicinal dihaloalkanes whose reaction with sulphydryl groups generates a further reactive alkylating species) (Van Bladeren et al., 1979; Guengerich et al., 1980). The measurement of protein adducts does, however, possess considerable advantages in comparison to that of DNA adducts owing to the ready availability of proteins from blood samples. Both hemoglobin (Hb) and, to a lesser extent, albumin (Ab) have been used for biomonitoring. Adducts in proteins are generally stable, in many cases up to a lifetime of the protein, which is approximately 120 days for Hb and $t^1/_2$ approximately 20 days for Ab.

The analytical methods employed for determining protein adducts are immunoassay and mass spectrometry. Modern mass spectrometric techniques are capable not only of characterizing intact adducted proteins, but also of determining the site in the sequence where the adduct is present. For example, electrospray ionization mass spectrometry has recently been used by Springer et al. (1993) to detect the acrylamide adduct with the β-chain of Hb (mass increase 71 Da with respect to the normal protein chain). Proteolytic digestion and subsequent mass spectrometry suggested the sites of adducts to be cysteine and N-terminal valine. Liquid secondary ion tandem mass spectrometry was used by Kaur et al. (1989) to study the covalent interactions between styrene oxide and human Hb. Tryptic peptides that had been modified by the epoxide were identified and then sequenced by collision-induced decomposition in the mass spectrometer to identify the amino acids where adducts had been formed (histidine and cysteine). Other examples where mass spectrometry and other physicochemical methods (e.g., nuclear magnetic resonance, X-ray crystallography) have elucidated the structures of adducts are for those formed between aflatoxin B_1 and lysine in Ab (Sabbioni et al., 1987), and aromatic amines (e.g., 4-aminobiphenyl) and β-93 cysteine in human Hb (Ringe et al., 1988). Further examples and details are reviewed in Tannenbaum et al. (1993). Such *in vitro* methods give valuable information regarding the nature of the adducts which one should consider biomonitoring in humans. However, the mass spectrometric methods described above, although highly sophisticated, do not have the sensitivity for monitoring the very low levels of adducted proteins in human samples. To quantitate such *in vivo* protein adducts by mass spectrometry, it is necessary first to digest or chemically degrade the protein to small molecules which may be analyzed by GC-MS or LC-MS.

The structural evidence obtained to date shows that the main sites of adduct formation in Hb are, as expected from their nucleophilicity, cysteine, histidine, N-terminal amino group (valine on both the α- and β-chain of Hb), and the carboxylic groups, in aspartic and glutamic acid. The main application of Ab adducts for biomonitoring has been for exposures to aflatoxin B_1, for which the main adducted site is lysine (Sabbioni et al., 1987), and to benzo(a)pyrene (Autrup et al., 1991), whose interaction products include acidic amino acids and histidine (Day et al., 1991).

A. ALBUMIN-CARCINOGEN ADDUCTS

Both immunological techniques and HPLC with fluoresence detection have been used to biomonitor aflatoxin B_1-albumin adducts. In brief, the method involves isolation of albumin from serum or plasma, followed by proteolytic digestion. The product is partially purified by reverse phase cartridge chromatography and quantitated by competitive ELISA (detection limit 1 pmol), or is more rigorously purified on an immunoaffinity column prior to HPLC-fluoresence (Wild et al., 1990). Widespread epidemiological studies have now been carried out, demonstrating the high incidence of adducts in some African and Asian populations (Wild et al., 1993).

The interaction of the benzo(a)pyrene metabolite BPDE with human serum albumin has been shown to generate carboxylate side chain esters and modified histidine residues (Day et al., 1991). As indicated in Table 6.2, the former adduct has potential as a biomarker of exposure, as it may be hydrolyzed to benzo(a)pyrene 7,8,9,10-tetrahydrotetrol, which may be analyzed with high sensitivity using GC-negative chemical ionization MS or immunoassay. For example, Autrup et al. (1991), using an antibody-recognizing BPDE-modified protein (Santella et al., 1986), demonstrated that human serum protein contained PAH adducts (range 0.01–0.76 fmol/µg protein), and that the levels were slightly higher in smokers compared to nonsmokers.

B. HEMOGLOBIN N-TERMINAL VALINE ADDUCTS

As indicated in Table 6.3, the N-terminal amino group in Hb has been demonstrated to form adducts with a variety of low molecular weight carcinogens. The analyses of these adducts were made possible by the development of a modified Edman degradation procedure by Törnqvist et al. (1986). This technique releases the adducted valine residue from the protein chain as a cyclic thiohydantoin derivative which may be extracted, purified, and quantified with selected ion monitoring GC-MS using a stable isotope-labeled internal standard. The sensitivity of these procedures has been reported to be in the range 1 to 10 pmol adduct/g Hb, which in the case of, say, ethylene oxide adducts, is comparable to the sensitivity achievable by the most sensitive DNA adduct analytical techniques. The ethylene oxide adduct N-(2-hydroxyethyl)valine has been the most widely studied. Agents which have been shown to contribute to increased levels of this adduct include 1-(2-chloroethyl)-1-nitrosourea cancer chemotherapeutic agents (Bailey et al., 1991), occupational ethylene oxide (Tates et al., 1991), cigarette smoke (Bailey et al., 1988), exhaust fumes (Törnqvist et al., 1988a), and endogenous sources of ethylene (Filser et al., 1992; Shen et al. 1989).

C. HEMOGLOBIN CYSTEINE ADDUCTS

Stable adducts at cysteine may be determined following total acidic hydrolysis of the globin. The resultant amino acids are subjected to chromato-

TABLE 6.3
Examples of Hemoglobin-Carcinogen Adducts

Amino acid	Adducted genotoxin
Cysteine	Aromatic amines (Stilwell et al. 1987)
	Methylating agents (Bailey et al. 1981)
	Acrylamide (Bailey et al. 1986; Bergmark et al. 1991)
	Styrene oxide (Ting et al. 1990)
	Acrylonitrile (Fennell et al. 1991)
	Benzene (Bechtold et al. 1992)
Aspartic acid, glutamic acid	Tobacco-specific nitrosamines (Murphy et al. 1990)
	Styrene oxide (Sepai et al. 1993)
	PAHs (Bechtold et al. 1991)
Histidine	Ethylene oxide (Osterman-Golkar et al. 1983)
	Propylene oxide (Osterman-Golkar et al. 1984)
N-terminal valine	Ethylene oxide (Bailey et al. 1988; Törnqvist et al. 1988a; Walker et al. 1992a)
	Propylene oxide (Törnqvist et al. 1988a)
	Styrene oxide (Christakopoulos et al. 1993)
	Acrylonitrile (Bergmark et al. 1993)
	Acrylamide (Bergmark et al. 1993)

graphy, derivatized, and quantitated by GC-MS. S-Methylcysteine was the first adduct studied in this way (Bailey et al., 1981), but the application of its analysis in humans was limited by high background levels (Table 6.4). Other stable cysteine adducts whose analysis may similarly be approached are those from acrylonitrile, acrylamide, and benzene (Table 6.3). A second approach to the analysis of cysteine thioether adducts is the use of Raney nickel, which, for example, has been shown to cleave, reductively, adducts of styrene oxide with cysteine in hemoglobin (or albumin) with the release of phenylethanol, which may then be determined by GC-MS (Rappaport et al., 1993).

Another class of cysteine carcinogen adducts are the sulphinamides, which are formed by aromatic amine metabolites. The amine is believed to be activated via P450 metabolism to an aryl hydroxylamine which reacts (maybe via the aryl nitroso compound) with cysteine sulphydryl groups. The sulphinamide products are not entirely stable and normally have shorter lifetimes than Hb. The analytical procedure for them involves their hydrolysis to the aromatic amine and the cysteic acid derivative using mild acidic or alkaline conditions. The amine is then extracted, derivatized, and subjected to GC-MS. Exceedingly sensitive assays for a variety of amines, e.g., 4-aminobiphenyl, have been developed, and human exposures to aromatic amines from smoking (Bryant et al., 1988) or occupational exposure (Bailey et al., 1990, and unpublished data) have been measured.

D. HEMOGLOBIN HISTIDINE ADDUCTS

Adducts are formed between electrophilic carcinogens and the N^τ or N^π of the imidazole ring of histidine. These may be identified and quantitated by GC-MS following acidic hydrolysis of the protein chain. This procedure, which is now not widely used, presents a considerable chromatographic challenge, as trace amounts of the adduct need to be separated from a vast excess of the normal amino acids. The applications of this method have been limited to adducts formed by low molecular weight alkylating agents and epoxides (Table 6.3). Studies of occupational exposure to ethylene oxide and propylene oxide revealed increased levels of the N^τ-histidine adducts of these epoxides (Farmer et al., 1986; Osterman-Golkar et al., 1984).

E. HEMOGLOBIN CARBOXYLIC ACID ADDUCTS

Esters of the carboxylic acids present in Hb (aspartic acid, glutamic acid, and C-terminal amino acids) have been shown to be formed following their interaction with epoxides formed from styrene and polycyclic aromatic hydrocarbons, and alkylating agents produced by metabolism of nitrosamines. These esters may be analyzed by procedures involving hydrolysis of the ester bond, extraction, purification of the released alcohol, derivatization, and GC-MS (Table 6.3). This is the basis of a particularly sensitive method for detecting human exposure to tobacco-specific nitrosamines (Hecht et al., 1993; Carmella et al., 1990). The alcohol analyzed in this case is 4-hydroxy-1-(3-pyridyl)-1-

TABLE 6.4
Approximate Background Levels of DNA and Hemoglobin-Carcinogen Adducts in Control Subjects

Backgrounds of Adducted Nucleic Acid Bases

Adduct	Modification/total nucleotides
N-7-Methylguanine[a]	$5/10^7$ (Mustonen et al. 1993)
N-7-(2-Hydroxyethyl)guanine[b]	$1/10^6$ (Cushnir et al. 1993)
8-Hydroxyguanine[c]	$8/10^6$ (Fraga et al. 1990)
cis-thymine glycol[d]	$3/10^6$ (Farmer et al. unpublished data)

Backgrounds of Adducted Amino Acids in Human Hemoglobin

Adduct	nmol adduct/g globin
S-Methylcysteine	16.4 (Bailey et al. 1981)
4-Aminobiphenyl-cysteine sulphinamide	0.00017 (Bryant et al. 1987; Perera et al. 1987)
N-Methylvaline	0.5 (Törnqvist et al. 1988b)
N-(2-Hydroxyethyl)valine	0.05 (Bailey et al. 1988)
N-(2-Cyanoethyl)valine	0-0.04 (Fennell et al. 1993)
N-(2-Hydroxyethyl)histidine	1.6 (Van Sittert et al. 1985)

[a]   Human bronchial DNA from nonsmokers
[b]   Human urine, assuming the source being 3g total DNA
[c]   24-month rat liver
[d]   Rat liver

butanone, which is converted to its pentafluorobenzoate derivative prior to GC-negative chemical ionization MS.

IV. DOSE-RESPONSE RELATIONSHIP FOR HEMOGLOBIN ADDUCTS AND DNA ADDUCTS

Dose response relationships have been determined for adduct formation by over 20 carcinogens with Hb (see review by Farmer, 1993). In most cases, the lower end of the dose response curve appears linear, but with high doses of carcinogen some deviations from linearity are observed. Some compounds at high concentrations show a greater than linear response, e.g., acrylamide, acrylonitrile, N-nitrosodimethylamine. This could be due, for example, to saturation of a detoxifying pathway at high doses. Other compounds show a lower than linear response at high doses (e.g., benzene, benzo(a)pyrene), which could be caused by saturation of activation pathways. Studies have similarly been carried out on dose-response relationships for the formation of some DNA-adducts, with generally similar conclusions. Positive deviations from linear dose-response relationships were seen with ethylmethane sulphonate and ethylene oxide, and negative deviations with 4-aminobiphenyl, N-nitrosodiethylamine, and 4-methylnitrosamino-1-(3-pyridyl)-1-butanone. (It is of interest to note that "upward deviations" seem to be mostly associated with directly acting compounds and "downward deviations" with compounds that are activated by metabolism.)

The overall conclusion from the dose response data available is that Hb adducts quantitatively reflect DNA adducts, particularly at low exposure doses. However, as demonstrated elegantly by Walker et al. (1992a, b, 1993) using ethylene oxide, the ratio between Hb and DNA adducts depends on length of exposure, interval since exposure, species, and tissue.

V. BACKGROUND LEVELS OF ADDUCTS

For many cases, DNA or Hb adducts have been detected in "control," supposedly unexposed, populations. The presence of these background adducts (summarized in Table 6.4), presumably reflects either endogenous exposure to the adduct-forming chemical (or to an endogenous compound of the same structure as the adduct) or exogenous exposure to the carcinogen (or the adduct) from an unknown source. Examples of all these categories are known. Thus, background levels of the ethylene oxide adduct N-(2-hydroxyethyl)valine in Hb are believed to be in part due to endogenous ethylene oxide (derived from endogenous ethylene) (Filser et al., 1992). Urinalysis of 7-methylguanine, as an indicator of DNA methylation, is limited by the presence of background levels of this purine, which is due in part to breakdown of tRNA which contains the base (Hall, 1971). Hb contains background levels of the 4-aminobiphenyl cysteine adduct (Bryant et al., 1987) and the ethylene oxide valine adduct (Bailey et al., 1988), which are thought to be due, respectively, to exogenous exposure to the amine or (in part) to the epoxide from environmental sources. Finally, some of the adduct 3-methyladenine in control urine is due to ingestion of the preformed compound in the food (Prevost et al., 1990). The presence of background levels of adducts suggests the existence of a background genetic risk, whose quantitation would seem to be a current research priority.

VI. CONCLUSION

Analytical methods exist now for the detection of DNA and protein adducts, both at the levels at which they are produced from environmental exposure to carcinogens, and also in control samples. The concentration of these adducts may be as low as 1 modified base per 10^8 normal DNA bases. The relationships between adducts and exposure are now quite well validated in animal experiments. For human exposure, the determination of adducts may additionally give information regarding genetic susceptibility, e.g., through polymorphism of a carcinogen-activating enzyme, of the exposed individual. The relationship between adducts and carcinogenic risk has also been investigated for several compounds in experimental animals.

It seems, therefore, that the stage is set for the quantitative use of adducts as a human carcinogenic risk monitor. However, there is great potential danger in equating numbers of adducts with carcinogenic risk, as adduct formation is only the first stage in the multi-step carcinogenesis process. Furthermore, each chemically different adduct will have a different "effectiveness" as a mutation inducer. Thus, measurement of adducts alone may be misleading, the more appropriate index of cancer-initiating exposure being "number of adducts × mutational effectiveness." Such a relationship is already used for assessing toxic risk from mixed dioxin exposures, which may be calculated from the sum of "exposure × toxicity" for all the components present (Safe, 1993). Although further work would need to be done before such an approach could be utilized, it should be emphasized that quantitative determination of adducts is already giving valuable information regarding the hazards associated with human exposure to carcinogens. Effectively, adduct measurement gives the biologically effective dose received of the carcinogen, which is a much more meaningful monitor of carcinogenic risk than external exposure measurements.

REFERENCES

Autrup, H., Seremet, T., and Sherson, D., Quantitation of polycyclic aromatic hydrocarbon-serum protein and lymphocyte DNA adducts in Danish foundry workers, in *Human Carcinogen Exposure. Biomonitoring and Risk Assessment*, Garner, R. C., Farmer, P. B., Steel, G. T., and Wright, A. S., Eds., IRC Press, Oxford, 1991, chap. 14.

Autrup, H., Seremet, T., Wakhisi, J., and Wasunna, A., Aflatoxin exposure measured by urinary excretion of aflatoxin B1-guanine adduct and hepatitis B infection in areas with different liver cancer incidence in Kenya, *Cancer Res.,* 47, 3430, 1987.

Bailey, E., Brooks, A. G., Bird, I., Farmer, P. B., and Street, B., Monitoring exposure to 4,4′-methylenedianiline by the gas chromatography-mass spectrometry determination of adducts to hemoglobin, *Anal. Biochem.,* 190, 175, 1990.

Bailey, E., Brooks, A. G. F., Dollery, C. T., Farmer, P. B., Passingham, B. J., Sleightholm, M. A., and Yates, D. W., Hydroxyethylvaline adduct formation in haemoglobin as a biological marker of cigarette smoke intake, *Arch. Toxicol.,* 62, 247, 1988.

Bailey, E., Connors, T. A., Farmer, P. B., Gorf, S. M., and Rickard, J., Methylation of cysteine in hemoglobin following exposure to methylating agents, *Cancer Res.,* 41, 2514, 1981.

Bailey, E., Farmer, P. B., Bird, I., Lamb, J. H., and Peal, J. A., Monitoring exposure to acrylamide by the determination of S-(2-carboxyethyl)cysteine in hydrolyzed hemoglobin by gas chromatography-mass spectrometry, *Anal. Biochem.,* 157, 241, 1986.

Bailey, E., Farmer, P. B., Tang, Y.-S., Vangikar, H., Gray, A., Slee, D., Ings, R. M. J., Campbell, D. B., McVie, J. G., and Dubbelman, R., Hydroxyethylation of hemoglobin by 1-(2-chloroethyl)-1-nitrosoureas, *Chem. Res. Toxicol.,* 4, 462, 1991.

Beach, A. C. and Gupta, R. C., Human biomonitoring and the ^{32}P-postlabeling assay, *Carcinogenesis,* 13, 1053, 1992.

Bechtold, W. E., Sun, J. D., Birnbaum, L. S., Yin, S. N., Guilan, L. L., Kasicki, S., Lucier, G., and Henderson, R. F., S-Phenylcysteine formation in hemoglobin as a biological exposure index for benzene, *Arch. Toxicol.,* 66, 303, 1992.

Bechtold, W. E., Sun, J. D., Wolff, R. K., Griffith, W. S., Kilmer, J. W., and Bond, J. A., Globin adducts of benzo(a)pyrene: markers of inhalation exposure as measured in F344/N rats, *J. Appl. Toxicol.*, 11, 115, 1991.

Bergmark, E., Calleman, C. J., and Costa, L. G., Formation of hemoglobin adducts of acrylamide and its epoxide metabolite glycidamide in the rat, *Toxic. Appl. Pharm.*, 111, 352, 1991.

Bergmark, E., Calleman, C. J., He, F., and Costa, L. G., Determination of hemoglobin adducts in humans occupationally exposed to acrylamide, *Toxicol. Appl. Toxicol.*, 120, 45, 1993.

Bryant, M. S., Skipper, P. L., Tannenbaum, S. R., and Maclure, M., Hemoglobin adducts of 4-aminobiphenyl in smokers and non-smokers, *Cancer Res.*, 47, 602, 1987.

Bryant, M. S., Vineis, P., Skipper, P. L., and Tannenbaum, S. R., Hemoglobin adducts of aromatic amines: associations with smoking status and type of tobacco, *Proc. Natl. Acad. Sci. U.S.A.*, 85, 9788, 1988.

Buchet, J. P., Gennart, J. P., Mercado-Calderon, F., Delavignette, J. P., Cupers, L., and Lauwerys, R., Evaluation of exposure to polycyclic aromatic hydrocarbons in a coke production and a graphite electrode manufacturing plant: assessment of urinary excretion of 1-hydroxypyrene as a biological indicator of exposure, *Brit. J. Ind. Med.*, 49, 761, 1992.

Burns, P. A., Gordon, A. J. E., and Glickman, B. W., Influence of neighboring base sequences on N-methyl-N-nitrosoguanidine mutagenesis in the lac-1 gene of Escherichia coli, *J. Mol. Biol.*, 194, 385, 1987.

Carmella, S. G., Kagan, S. S., Kagan, M., Foiles, P. G., Palladino, P. G., Quart, A. M., Quart, E., and Hecht, S. S., Mass spectrometric analysis of tobacco-specific nitrosamine hemoglobin adducts in snuff-dippers, smokers, and non-smokers, *Cancer Res.*, 50, 5438, 1990.

Christakopoulos, A., Bergmark, E., Zorcec, V., Norppa, H., Maki-Paakkanen, J., and Osterman-Golkar, S., Monitoring occupational exposure to styrene from hemoglobin adducts and metabolites in blood, *Scand. J. Work Environ. Health*, 19, 255, 1993.

Cocker, J., Boobis, A., and Davies, D. S., Determination of the N-acetyl metabolites of 4,4'-methylene dianiline and 4,4'-methylene-bis(2-chloroaniline) in urine, *Biomed. Mass Spectrom.*, 17, 161, 1988.

Cooper, D. P., Griffin, K. A., and Povey, A. C., Immunoaffinity purification combined with ^{32}P-postlabeling for the detection of O6-methylguanine in DNA from human tissues, *Carcinogenesis*, 13, 469, 1992.

Cosman, M., de los Santos, C., Fiala, R., Hingerty, B. E., Ibanez, V., Luna, E., Harvey, R., Geacintov, N. E., Broyde, S., and Patel, D. J., Solution conformation of the (+)-cis-anti-(BP)-dG adduct in a DNA duplex:intercalation of the covalently attached benzo(a)pyrenyl ring into the helix and displacement of the modified deoxyguanosine, *Biochemistry*, 32, 4145, 1993.

Cushnir, J. R., Naylor, S., Lamb, J. H., and Farmer, P. B, Tandem mass spectrometric approaches for the analysis of alkylguanines in human urine, *Org. Mass Spectrom.*, 28, 552, 1993.

Day, B. W., Skipper, P. L., Zaia, J., and Tannenbaum, S. R., Benzo(a)pyrene anti diol epoxide covalently modifies human serum albumin carboxylate side chains and imidazole side chain of histidine, *J. Amer. Chem. Soc.*, 113, 8505, 1991.

Den Engelse, L., Menkveld, G. J., De Brij, R.-J., and Tates, A. D., Formation and stability of alkylated pyrimidines and purines (including imidazole ring-opened 7-alkylguanine) and alkylphosphotriesters in liver DNA of adult rats treated with ethylnitrosourea or dimethylnitrosamine, *Carcinogenesis*, 7, 393, 1986.

Dizdaroglu, M., Characterisation of free radical induced damage to DNA by the combined use of enzymatic hydrolysis and gas chromatography-mass spectrometry, *J. Chromatogr.*, 367, 357, 1986.

Doll, R. and Peto, R., The causes of cancer: Quantitative estimates of avoidable risks of cancer in the United States today, *J. Nat. Cancer Inst.*, 66, 1191, 1981.

Eberle, G., Glüsenkamp, K. H., Drosdziok, W., and Rajewsky, M. F., Monoclonal antibodies for the specific detection of 3-alkyladenines in nucleic acids and body fluids, *Carcinogenesis*, 11, 1753, 1990.

Engström, K., Harkonen, H., Pekari, K., and Rantanen, J., Evaluation of occupational styrene exposure by ambient air and urine analysis, *Scan. J. Work Environ. Health*, 4, Suppl. 2, 121, 1978.

Everson, R. B., Randerath, E., Santella, R. M., Cefalo, R. C., Avitts, T. A., and Randerath, K., Detection of smoking-related covalent adducts in human placenta, *Nature*, 231, 54, 1986.

Farmer, P. B., Biomarkers as molecular dosimeters of genotoxic substances, in *Use of Biomarkers in Assessing Health and Environmental Impacts of Chemicals*, Travis, C. C., Ed., Plenum Press, 1993, 53.

Farmer, P. B., Bailey, E., Gorf, S. M., Törnqvist, M., Osterman-Golkar, S., Kautiainen, A., and Lewis-Enright, D. P., Monitoring human exposure to ethylene oxide by the determination of haemoglobin adducts using gas chromatography-mass spectrometry, *Carcinogenesis*, 7, 637, 1986.

Fedtke, N., Boucheron, J. A., Turner, M. J., and Swenberg, J. A., Vinyl chloride induced adducts. I: Quantitative determination of N2,3-ethenoguanine based on electrophore labeling, *Carcinogenesis*, 11, 1279, 1990.

Fedtke, N. and Swenberg, J. A., *Quantitative Analysis of DNA Adducts: the Potential for Mass Spectrometric Techniques,* Groopman, J. D. and Skipper, P. L., Eds., CRC Press, Boca Raton, FL, 1991, chap. 10.

Felton, J. S., Turteltaub, K. W., Vogel, J. S., Balhorn, R., Gledhill, B. L., Southon, J. R., Caffee, M. W., Finkel, R. C., Nelson, D. E., Proctor, I. D., and Davis, J. C., Accelerator mass spectrometry in the biomedical sciences: applications in low-exposure biomedical and environmental dosimetry, *Nucl. Inst. and Meth. in Phys. Res.,* B52, 517, 1990.

Fennell, T. R., Calleman, C. J., Costa, L. G., and MacNeela, J. P., N-(2-cyanoethyl)valine in hemoglobin from workers exposed to acrylonitrile, *Proc. Amer. Assoc. Cancer Res.,* 34, 153, 1993.

Fennell, T. R., MacNeela, J. P., Turner, M. J., and Swenberg, J. A., Haemoglobin adduct formation by acrylonitrile in rats and mice, in *Human Carcinogen Exposure. Biomonitoring and Risk Assessment*, Garner, R. C., Farmer, P. B., Steel, G. T., and Wright, A. S., Eds., IRC Press, Oxford, 1991, chap. 18.

Filser, J. G., Denk, B., Törnqvist, M., Kessler, W., and Ehrenberg, L., Pharmacokinetics of ethylene in man; body burden with ethylene oxide and hydroxyethylation of hemoglobin due to endogenous and environmental ethylene, *Arch. Toxicol.,* 66, 157, 1992.

Fraga, C. G., Shigenaga, M. K., Park, J.-W., Degan, P., and Ames, B. N., Oxidative damage to DNA during aging: 8-hydroxy-2'-deoxyguanosine in rat organ DNA and urine, *Proc. Natl. Acad. Sci. U.S.A.,* 87, 4533, 1990.

Friesen, M. D., Garren, L., Prevost, V., and Shuker, D. E. G., Isolation of urinary 3-methyladenine using immunoaffinity columns prior to determination by low-resolution gas chromatography-mass spectrometry, *Chem. Res. Toxicol.,* 4, 102, 1991.

Glickman, B. W., Horsfall, M. J., Gordon, A. J. E., and Burns, P. A., Nearest neighbor affects G:C to A:T transitions induced by alkylating agents, *Environ. Health Perspec.,* 76, 29, 1987.

Groopman, J. D., Hasler, J., Trudel, L. J., Pikul, A., Donahue, P. R., and Wogan, G. N., Molecular dosimetry in rat urine of aflatoxin-N7-guanine and other aflatoxin metabolites by multiple monoclonal antibody affinity chromatography and high performance liquid chromatography, *Cancer Res.,* 52, 267, 1992a.

Groopman, J. D., Trudel, L. J., Donahue, P. R., Marshak-Rothstein, A., and Wogan, G. N., High affinity monoclonal antibodies for aflatoxins and their application to solid phase immunoassays, *Proc. Natl. Acad. Sci. U.S.A.,* 81, 7728, 1984.

Groopman, J. D., Zhu, J., Donahue, P. R., Pikul, A., Zhang, L., Chen, J.-S., and Wogan, G. N., Molecular dosimetry of urinary aflatoxin DNA adducts in people living in Guangxi autonomous region, Peoples Republic of China, *Cancer Res.,* 52, 45, 1992b.

Guengerich, F. P., Crawford, W. M., Domoradzki, J. Y., MacDonald, T. L., and Watanabe, P. G., *In vitro* activation of 1,2-dichloroethane by microsomal and cytosolic enzymes, *Toxicol. Appl. Pharmacol.,* 55, 303, 1980.

Hall, R., *The Modified Nucleosides in Nucleic Acids*, Columbia Univ. Press, New York, 1971.

Harris, C. C., Vahakangas, K., Newman, M. J., Trivers, G. E., Shamsuddin, A., Sinopoli, N., Mann, D. L., and Wright, W. E., Detection of benzo(a)pyrene diol epoxide-DNA adducts in peripheral blood lymphocytes and antibodies to the adducts in serum from coke oven workers, *Proc. Natl. Acad. Sci. U.S.A.*, 82, 6672, 1985.

Hartley, J. A., Lown, J. W., Mattes, W. B., and Kohn, K. W., DNA sequence specificity of antitumour agents. Oncogenes as possible targets for cancer therapy, *Acta Oncologia*, 27, 503, 1988.

Hecht, S. S., Carmella, S. G., Foiles, P. G., Murphy, S. E., and Peterson, L. A., Tobacco-specific nitrosamine adducts: studies in laboratory animals and humans, *Environ. Health Perspec.*, 99, 57, 1993.

Jongeneelen, F. J., Bos, R. P., Anzion, R. B. M., Theuws, J. J. G., and Henderson, P. T., Biological monitoring of polycyclic aromatic hydrocarbons. Metabolites in urine, *Scan. J. Work Environ. Health*, 12, 137, 1986.

Kaur, S., Hollander, D., Haas, R., and Burlingame, A. L., Characterization of structural xenobiotic modifications in proteins by high sensitivity tandem mass spectrometry, *J. Biol. Chem.*, 264, 16981, 1989.

Kaur, S., Pongracz, K., Bodell, W. J., and Burlingame, A. L., Bis(hydroxyphenylethyl)deoxy-guanosine adducts identified by (^{32}P)-postlabeling and four-sector tandem mass spectrometry: unanticipated adducts formed upon treatment of DNA with styrene 7,8-oxide, *Chem. Res. Toxicol.*, 6, 125, 1993.

Kriek, E., van Schooten, F. J., Hillebrand, M. J. X., van Leeuwen, F. E., den Engelse, L., de Looff, A. J. A., and Dijkmans, A. P. G., DNA adducts as a measure of lung cancer risk in humans exposed to polycyclic aromatic hydrocarbons, *Environ. Health Perspec.*, 99, 71, 1993.

Kruger, M. and Salomon, G. Z., Die Alloxurbasen des Harnes, *Physiol. Chem.*, 24, 364, 1898.

Lawley, P. D., DNA as a target of alkylating carcinogens, *Brit. Med. Bull.*, 36, 19, 1980.

Manchester, D. K., Weston, A., Choi, J.-S., Trivers, G. E., Fennessey, P. V., Quintana, E., Farmer, P. B., Mann, D., and Harris, C. C., Detection of benzo(a)pyrene diol epoxide-DNA adducts in human placenta, *Proc. Nat. Acad. Sci. U.S.A.*, 85, 9243, 1988.

Mattes, W. B., Hartley, J. A., Kohn, K. W., and Matheson, D. W., GC-rich regions in genomes as targets for DNA alkylation, *Carcinogenesis*, 9, 2065, 1988.

Mumford, J. L., Lee, X., Lewtas, J., Young, T. L., and Santella, R. M., DNA adducts as biomarkers for assessing exposure to polycyclic aromatic hydrocarbons in tissues from Xuan Wei women with high exposure to coal combustion emissions and high lung cancer mortality, *Environ. Health Perspec.*, 99, 83, 1993.

Murphy, S. E., Palomino, A., Hecht, S. S., and Hoffmann, D., Dose-response study of DNA and hemoglobin adduct formation by 4-(methylnitrosamino)-1-(3-pyridyl)-1-butanone in F344 rats, *Cancer Res.*, 50, 5446, 1990.

Mustonen, R., Schoket, B., and Hemminki, K., Smoking-related DNA adducts: ^{32}P-postlabeling analysis of 7-methylguanine in human bronchial and lymphocyte DNA, *Carcinogenesis*, 14, 151, 1993.

Osterman-Golkar, S., Bailey, E., Farmer, P. B., Gorf, S. M., and Lamb, J. H., Monitoring exposure to propylene oxide through the determination of hemoglobin alkylation, *Scand. J. Work Environ. Health*, 10, 99, 1984.

Osterman-Golkar, S., Farmer, P. B., Segerback, D., Bailey, E., Calleman, C. J., Svensson, K., and Ehrenberg, L., Dosimetry of ethylene oxide in the rat by quantitation of alkylated histidine in hemoglobin, *Teratogen. Carcinogen. Mutagen.*, 3, 395, 1983.

Perera, F. P., Hemminki, K., Gryzbowska, E., Motykiewicz, G., Michalska, J., Santella, R. M., Young, T.-L., Dickey, C., Brandt-Rauf, P., DeVivo, I., Blaner, W., Tsai, W.-Y., and Chorazy, M., Molecular and genetic damage in humans from environmental pollution in Poland, *Nature*, 360, 256, 1992.

Perera, F. P., Santella, R. M., Brenner, D., Poirier, M. C., Munshi, A. A., Fischman, H. K. and Van Ryzin, J., DNA adducts, protein adducts, and sister chromatid exchange in cigarette smokers and non-smokers, *J. Natl. Cancer Inst.*, 79, 449, 1987.

Poirier, M. C., Immunochemical methods for assaying carcinogen-DNA adducts, in *Human Carcinogen Exposure. Biomonitoring and Risk Assessmment*, Garner, R. C., Farmer, P. B., Steel, G. T., and Wright, A. S., Eds., IRC Press, Oxford, 1991, chap 5.

Poirier, M. C., Reed, E., Shamkhani, H., Tarone, R. E., and Gupta-Burt, S., Platinum drug-DNA interactions in human tissues measured by cisplatin-DNA enzyme-linked immunosorbent assay and atomic absorbance spectroscopy, *Environ. Health Perspec.*, 99, 149, 1993.

Prevost, V. and Shuker, D. E. G., Urinary 3-alkyladenines as markers of tobacco-smoke exposure in humans, *Human Toxicol.*, 11, 427, 1992.

Prevost, V., Shuker, D. E. G., Bartsch, H., Pastorelli, R., Stillwell, W. G., Trudel, L. J., and Tannenbaum, S. R., The determination of urinary 3-methyladenine by immunoaffinity chromatography-monoclonal antibody-based ELISA: use in human biomonitoring studies, *Carcinogenesis*, 11, 1747, 1990.

Prevost, V., Shuker, D. E. G., Friesen, M. D., Eberle, G., Rajewsky, M. F., and Bartsch, H., Immunoaffinity purification and gas chromatography-mass spectrometric quantification of 3-alkyladenines in urine: metabolism studies and basal excretion levels in man, *Carcinogenesis*, 14, 199, 1993.

Randerath, K., Reddy, M. V., and Gupta, R. C., ^{32}P-Postlabelling test for DNA damage, *Proc. Natl. Acad. Sci. U.S.A.*, 78, 6126, 1981.

Rappaport, S. M., Ting, D., Jin, Z., Yeowell-O'Connell, K., Waidyanatha, S., and McDonald, T., Application of Raney nickel to measure adducts of styrene oxide with hemoglobin and albumin, *Chem. Res. Toxicol.*, 6, 238, 1993.

Rees, N. M. and Tennant, D. R., in *Safety of Chemicals in Food- Chemical Contaminants*, Watson, D. H., Ed., Ellis Horwood, 1993, 157.

Ringe, D., Turesky, R. J., Skipper, P . L., and Tannenbaum, S. R., Structure of the single stable hemoglobin adduct formed by 4-aminobiphenyl *in vivo*, *Chem. Res. Toxicol.*, 1, 22, 1988.

Ross, R. K., Yuan, J.-M., Yu, M. C., Wogan, G. N., Qian, G.-S., Tu, J.-T., Groopman, J. D., Gao, Y.-T., and Henderson, B. E., Urinary biomarkers and risk of hepatocellular carcinoma, *Lancet*, 339, 943, 1992.

Sabbioni, G., Skipper, P. L., Buchi, G., and Tannenbaum, S. R., Isolation and characterisation of the major serum albumin adduct formed by aflatoxin B$_1$ *in vivo* in rats, *Carcinogenesis*, 8, 819, 1987.

Sabbioni, G., Tannenbaum, S. R., and Shuker, D. E. G., Synthesis of volatile, fluorescent 7-methylguanine derivatives via reaction with 2-substituted fluorinated malondialdehydes, *J. Org. Chem.*, 51, 3244, 1986.

Safe, S., Development of bioassays and approaches for the risk assessment of 2,3,7,8-tetrachlorodibenzo-p-dioxin and related compounds, *Environ. Health Perspec.*, 101, Suppl. 3, 317, 1993.

Santella, R. M., Lin, C. D., and Dharmaraja, N., Monoclonal antibodies to a benzo(a)pyrene diolepoxide modified protein, *Carcinogenesis*, 7, 441, 1986.

Sepai, O., Anderson, D., Street, B., Bird, I., Farmer, P. B., and Bailey, E., Monitoring of exposure to styrene oxide by GC-MS analysis of phenylhydroxyethyl esters in hemoglobin, *Arch. Toxicol.*, 67, 28, 1993.

Shen, J., Kessler, W., Denk, B., and Filser, J. G., Metabolism and endogenous production of ethylene in rat and man, *Arch. Toxicol. Suppl.*, 13, 237, 1989.

Shooter, K. V., DNA phosphotriesters as indicators of cumulative carcinogen-induced damage, *Nature*, 274, 612, 1978.

Shuker, D. E. G., Determination of 7-methylguanine by immunoassay, in *Methods for Detecting DNA Damaging Agents in Humans: Applications in Cancer Epidemiology and Prevention*, Bartsch, H., Hemminki, K., and O'Neill, I. K., Eds., IARC Scientific Publications 89, International Agency for Research on Cancer, Lyons, France, 1988, 296.

Shuker, D. E. G. and Durand, M. J., Detection of N-7-alkylguanines by fluorescent labeling and immunochemical methods, *Proc. Am. Assoc. Cancer Res.*, 33, 146, 1992.

Shuker, D. E. G. and Farmer, P. B., Relevance of urinary DNA adducts as markers of carcinogen exposure, *Chem. Res. Toxicol.*, 5, 450, 1992.

Springer, D. J, Bull, R. J., Goheen, S. C., Sylvester, D. M., and Edmonds, C. G., Electrospray ionization mass spectrometric characterization of acrylamide adducts to hemoglobin, *J. Toxicol. Environ. Health*, 40, 161, 1993.

Stillwell, W. G., Bryant, M. S., and Wishnok, J. S., GC/MS analysis of biologically important aromatic amines. Application to human dosimetry, *Biomed. Environ. Mass Spectrom.*, 14, 221, 1987.

Tannenbaum, S. R., Skipper, P. L., Wishnok, J. S., Stillwell, W. G., Day, B. W., and Taghizadeh, K., Characterization of various classes of protein adducts, *Environ. Health Perspec.*, 99, 51, 1993.

Tates, A. D., Grummt, T., Törnqvist, M., Farmer, P. B., van Dam, F. J., van Mossel, H., Schoemaker, H. M., Osterman-Golkar, S., Uebel, Ch., Tang, Y.-S., Zwinderman, A. H., Natarajan, A. T., and Ehrenberg, L., Biological and chemical monitoring of occupational exposure to ethylene oxide, *Mut. Res.*, 250, 483, 1991.

Ting, D., Smith, M. T., Doane-Setzer, P., and Rappaport, S. M., Analysis of styrene oxide-globin adducts based upon reaction with Raney nickel, *Carcinogenesis*, 11, 755, 1990.

Tomatis, L., Aitio, A., Wilbourn, J., and Shuker, L., Human carcinogens so far identified, *Jpn. J. Cancer Res.*, 80, 795, 1989.

Törnqvist, M., Kautiainen, A., Gatz, R. N., and Ehrenberg, L., Hemoglobin adducts in animals exposed to gasoline and diesel exhausts, 1. Alkenes, *J. Appl. Toxicol.*, 8, 159, 1988a.

Törnqvist, M., Mowrer, J., Jensen, S., and Ehrenberg, L., Monitoring of environmental cancer initiators through hemoglobin adducts by a modified Edman Degradation method, *Anal. Biochem.*, 154, 255, 1986.

Törnqvist, M., Osterman-Golkar, S., Kautiainen, A., Naslund, M., Calleman, C. J., and Ehrenberg, L., Methylations in human hemoglobin, *Mut. Res.*, 204, 521, 1988b.

Turteltaub, K. W., Felton, J. S., Gledhill, B. L., Vogel, J. S., Southon, J. R., Caffee, M. W., Finkel, R. C., Nelson, D. E., Proctor, I. D., and Davis, J. C., Accelerator mass spectrometry in biological dosimetry: relationship between low-level exposure and covalent binding of heterocyclic amine carcinogens to DNA, *Proc. Natl. Acad. Sci. U.S.A.*, 87, 5288, 1990.

Van Bladeren, P. J., Van der Gen, A., Breimer, D. D., and Mohn, G. R., Stereoselective activation of vicinal dihalogen compounds to mutagens by glutathione conjugation, *Biochem. Pharmacol.*, 28, 2521, 1979.

Van Delft, J. H. M., van Winden, M. J. M., van den Ende, A. M. C., and Baan, R. A., Determining N7-alkylguanine adducts by immunochemical methods and HPLC with electrochemical detection: applications in animal studies and in monitoring human exposure to alkylating agents, *Environ. Health Perspec.*, 99, 25, 1993.

Van Sittert, N. J., de Jong, G., Clare, M. G., Davies, R., Dean, B. J., Wren, L. J., and Wright, A. S., Cytogenetic, immunologic, and haematological effects in workers in an ethylene oxide manufacturing plant, *Brit. J. Ind. Med.*, 42, 19, 1985.

Vodicka, P., Vodickova, L., and Hemminki, K., ^{32}P-Postlabeling of DNA adducts of styrene-exposed lamination workers, *Carcinogenesis*, 14, 2059, 1993.

Walker, V. E., Fennell, T. R., Upton, P. B., MacNeela, J. P., and Swenberg, J. A., Molecular dosimetry of DNA and hemoglobin adducts in mice and rats exposed to ethylene oxide, *Environ. Health Perspec.*, 99, 11, 1993.

Walker, V. E., Fennell, T. R., Upton, P. B., Skopek, T. R., and Swenberg, J. A., Molecular dosimetry of ethylene oxide. II: Formation and persistence of 7-(2-hydroxyethyl)guanine in DNA following repeated exposures of rats and mice, *Cancer Res.*, 52, 4328, 1992b.

Walker, V. E., MacNeela, P., Swenberg, J. A., Turner, M. J. Jr., and Fennell, T. R., Molecular dosimetry of ethylene oxide. I: Formation and persistence of N-(2-hydroxyethyl)valine in hemoglobin following repeated exposures of rats and mice, *Cancer Res.*, 52, 4320, 1992a.

Weston, A., Bowman, E. D., Shields, P. G., Trivers, G. E., Poirier, M. C., Santella, R. M., and Manchester, D. K., Detection of polycyclic aromatic hydrocarbon-DNA adducts in human lung, *Environ. Health Perspec.*, 99, 257, 1993.

Weston, A., Rowe, M. L., Manchester, D. K., Farmer, P. B., Mann, D. L., and Harris, C. C., Fluorescence and mass spectral evidence for the formation of benzo(a)pyrene anti-diol-epoxide-DNA and -hemoglobin adducts in humans, *Carcinogenesis*, 10, 251, 1989.

White, I. N., De Matteis, F., Davies, A., Smith, L. L., Crofton-Sleigh, C., Venitt, S., Hewer, A., and Phillips, D. H., Genotoxic potential of tamoxifen and analogues in female Fischer F344/n rats, DBA/2 and C57BL/6 mice and in human MCL-5 cells, *Carcinogenesis*, 13, 2197, 1992.

Wild, C., Jiang, Y., Sabbioni, G., Chapot, B., and Montesano, R., Evaluation of methodologies for quantitation of aflatoxin-albumin adducts and their application to human exposure assessment, *Cancer Res.,* 50, 245, 1990.

Wild, C. P., Jansen, L. A. M., Cova, L., and Montesano, R., Molecular dosimetry of aflatoxin exposure: contribution to understanding the multifactorial etiopathogenesis of primary hepatocellular carcinoma with particular reference to hepatitis B virus, *Environ. Health Perspec.,* 99, 115, 1993.

Zarbl, H., Sukumar, S., Arthur, A. V., Martin-Zanca, D., and Barbacid, M., Direct mutagenesis of Ha-ras-1 oncogenes by N-nitroso-N-methylurea during initiation of mammary carcinogenesis in rats, *Nature*, 315, 382, 1985.

Chapter 7

THE RELATIONSHIP OF MOLECULAR INTERACTIONS WITH DNA TO CANCER: SCIENTIFIC DEVELOPMENTS

P.N. Magee

CONTENTS

I. Introduction 125

II. Establishment of the Existence of Covalent Carcinogen-DNA
   Binding. A Brief Historical Comment 126

III.. Implications of Carcinogen-DNA Binding for Mechanistic
   Theories of Chemical Carcinogenesis 128

IV. Genotoxic and Nongenotoxic Carcinogens 132

V. Practical Aspects of DNA Adduct Formation *In Vivo* 133
   A. Implications of Genotoxicity for Safety Evaluation.
      The Case of Tamoxifen 133
   B. DNA Adducts as Markers of Exposure to Carcinogens 135
   C. DNA Adduct Dosimetry and Risk Assessment 137

VI. Conclusions 138

References 139

I. INTRODUCTION

The discovery by James and Elizabeth Miller, in 1947, that the carcinogenic dye *p*-dimethylaminoazobenzene became covalently bound to liver proteins of rats fed this compound was a landmark in chemical carcinogenesis investigation (Miller and Miller, 1947). The extension of these findings to aromatic amines, polycyclic hydrocarbons, and other types of carcinogens by these and other workers led to the important generalization by the Millers that some chemical carcinogens require metabolic activation to yield reactive electrophilic intermediates which form covalent bonds with nucleophilic sites in cellular constituents (Miller and Miller, 1966). The strong correlation between the presence of protein bound (Miller and Miller, 1966) carcinogen and the

0-8493-9229-2/95/$0.00+$.50
© 1995 by CRC Press Inc.

induction of cancer led to proposals of various hypotheses suggesting a causal relationship, particularly by workers at the University of Wisconsin, including the Millers, Potter, Heidelberger, and Pitot (Miller and Miller, 1959). At that time unequivocal evidence for binding of carcinogens to nucleic acids had not been obtained. However, the possibility that some carcinogens might act through binding to DNA had been suggested by Boyland in 1952 (Boyland, 1952).

During the intervening years, the majority opinion has come to favor DNA binding as the significant event in the initiation of chemical carcinogenesis, and the following short article will discuss briefly how this change came about and some of its theoretical and practical implications.

II. ESTABLISHMENT OF THE EXISTENCE OF COVALENT CARCINOGEN-DNA BINDING. A BRIEF HISTORICAL COMMENT

It is almost a commonplace at the present time that many of the most potent known chemical carcinogens react covalently with DNA in the organs of animals exposed to them. These interactions may cause somatic mutations in certain critical genes resulting in initiation of the process of carcinogenesis. Rosenkranz and Klopman summarized the position in 1990 as follows: "The somatic mutation theory of cancer achieved acceptance as a result of the pioneering studies of Miller and Miller (1977) and their associates who established that many carcinogens were metabolized to electrophilic intermediates capable of reacting with cellular DNA" (Rosenkranz and Klopman, 1990). This statement was chosen from among many similar ones because it correctly and justifiably pays tribute to the outstanding work of the Millers and their colleagues. However, in this context the word *capable* has special significance because it was not until considerably later than the discovery of carcinogen-protein interactions that covalent binding to nucleic acids was established and its significance appreciated. The views of the distinguished colleague of the Millers, Charles Heidelberger, expressed in 1964, and quoted as follows, clearly illustrate this: "It has been shown that some papers reporting *in vitro* interaction of carcinogenic hydrocarbons and DNA are in error, and that no binding had actually occurred. In studies *in vivo*, no radioactivity could be detected in DNA from mouse skin purified by cesium chloride density gradient centrifugation following application of highly radioactive hydrocarbons. A possible theory that could explain carcinogenesis by means of a derangement in metabolic control circuitry has been proposed. This theory eliminates the necessity of postulating a direct interaction between the carcinogen and the genetic material" (Heidelberger, 1964; Pitot and Heidelberger, 1963). Although Heidelberger comments elsewhere in the same paper that "there are indications, but by no means proof, that the alkylating agents induce cancer as a result of direct alkylation of DNA," it is clear that he totally rejected, at the time of writing, that the polycyclic hydrocarbons, a major group of chemical carcinogens, reacted with cellular DNA in animals exposed to them (Heidelberger, 1964).

The remainder of this section will provide a short discussion of some of the experimental work that led from the 1964 position of Heidelberger to the 1990 statement by Rosenkranz and Klopman quoted earlier, which is now the generally held view. Special emphasis will be placed on work done at the MRC Laboratories at Carshalton in the United Kingdom and much briefer mention made of the very important work of Peter Brookes and Philip Lawley of the Chester Beatty Institute in London.

The first published report of carcinogen binding to nucleic acids *in vivo* appears to be that of Wheeler and Skipper in 1957 who found binding of ^{14}C from injected nitrogen mustard-$^{14}CH_3$ to the nucleic acid fraction of animal tissues chiefly as a result of combination with nucleic acids (Wheeler and Skipper, 1957). Shortly afterwards Brookes and Lawley reported reaction of ^{35}S-labeled mustard gas with DNA and RNA *in vitro*, and with the nucleic acids of tobacco mosaic virus, *Bacillus megatherium,* and Erlich ascites tumor cells. The latter cells were recovered from mice bearing this tumor (Brookes and Lawley, 1960). In the same year, the writer and Emmanuel Farber demonstrated reaction of dimethylnitrosamine as well as ethionine, both hepatic carcinogens, with rat liver RNA *in vivo* (Farber and Magee, 1960). The contribution of the Carshalton workers to proving the existence of carcinogen-DNA interaction derived from the earlier discovery of the carcinogenic action of dimethylnitrosamine reported by Magee and Barnes in 1956 (Magee and Barnes, 1956).

Largely resulting from the extensive studies of Druckrey, Schmähl, Preussmann, and their colleagues (Druckrey et al., 1967) and later similar work on structure/activity relationships by Lijinsky and his associates (Lijinsky, 1992) the N-nitroso compounds are recognized to be a large group of chemical carcinogens of wide-ranging activity, that induce cancer in many animal species and in most organs of the body (Lijinsky, 1992; Preussmann and Stewart, 1984). There is also much circumstantial but suggestive evidence that these compounds, which occur in the environment, may contribute to the causation of some human cancer (Magee, 1989).

Unequivocal evidence for the formation of a carcinogen-DNA adduct by dimethylnitrosamine was reported in 1962 by Magee and Farber, who showed the presence of 7-methylguanine in the DNA and RNA of livers and kidneys of rats that had received a single dose of dimethylnitrosamine sufficient to induce kidney tumors. This demonstration was made virtually conclusive by the identification of the major chromatographic peak of radioactivity in hydrolysates of the nucleic acids from treated animals as 7-methylguanine on the basis of its identical UV spectra at different pHs with an authentic sample of the methylated base (Magee and Farber, 1962). This finding was subsequently confirmed by others using mass spectrometry.

At about the same time, covalent binding of radioactively labeled polycyclic aromatic hydrocarbons to nucleic acids of mouse skin was convincingly shown to occur by Brookes and Lawley, who also noted a relationship between their carcinogenic power and their binding to DNA (Brookes and Lawley, 1964). It was not until later that the identity of the diolepoxide DNA adducts of the polycyclic hydrocarbons was established by Sims, Grover, and their co-workers (Sims et al., 1974).

III. IMPLICATIONS OF CARCINOGEN-DNA BINDING FOR MECHANISTIC THEORIES OF CHEMICAL CARCINOGENESIS

Covalent binding of activated carcinogens with DNA *in vivo* has now been established for a large number of chemicals, and this has led to the perhaps rather uncritical acceptance of the proposition that such interactions are causally related to the subsequent development of cancer (Rosenkranz and Klopman, 1990). Since a number of chemicals appear to be carcinogenic without detectable DNA interaction and not all such interactions *in vivo* are followed by tumor formation, it seems that the postulated causal sequence should be examined more closely. In what follows, an attempt will be made to do this with reference to Koch's Postulates as recently "restructured into more general terms suitable for molecular pathology" by Hall and Lemoine (1991) as follows:

1. The identification of the cellular and molecular events common to a given pathological process.
2. The characterization of the temporal (and spatial) sequence of these events.
3. Demonstration that in some suitable model system, introduction of the observed molecular changes leads to an altered phenotype characteristic of the disease process.
4. Demonstration that correction of the molecular defect in a suitable model system rectifies the abnormalities characteristic of the disease process.

Carcinogenesis by N-nitroso compounds will be taken as a representative system for application of the restructured Koch's Postulates to the proposal that covalent DNA binding can be responsible for causing cancer. As described above, the first evidence of covalent DNA binding of these carcinogens was the presence of 7-methylguanine in the DNA of treated animals. However, it became apparent that this adduct was not responsible for their carcinogenic action because methylmethane sulfonate, which acts as an alkylating agent predominantly by the S_N2 mechanism, proved to be a relatively weak carcinogen in the rat, even though it yielded similar amounts of 7-methylguanine in the kidney DNA of treated animals as were induced by dimethylnitrosamine, which was powerfully carcinogenic (Swann and Magee, 1968). It was soon realized that reaction on the 7-position of guanine occurred to the greatest extent quantitatively, but that methylation also occurred at all other accessible nucleophilic sites on the DNA molecule. The important suggestion was made by Loveless in 1969 that the 0_6-position of guanine was most likely to be the crucial site of reaction of the methyldiazonium ion intermediates of the nitroso compounds (Loveless, 1969) leading later to the generalization that oxygen atoms rather than nitrogens were of greater importance in determining the mutagenic and carcinogenic actions of the alkylating agents (Singer, 1976). A major objection to the proposal that 0_6-methylguanine in DNA or any other methylated base could be the initial cause of tumors arose from the observation that, as expected, the levels of methylation with methylnitrosourea were much the same throughout the animal body (Swann and Magee, 1968) whereas the sites of tumor induction, depending on other conditions, were variable and sometimes involved one or only a few organs. A plausible explanation for this discrepancy was proposed by Goth and Rajewsky (1974) and by Margison and Kleihues (1975) who showed that 0_6-alkylgroups were much more persistent in the brain, where tumors were induced, than in the liver, where they were not. These findings suggested the existence of a DNA repair mechanism, presumably enzymic, which selectively nullified the effects of 0_6-methyl or 0_6-ethyl substituents while leaving the N-7 alkyl adducts unaffected. The probable nature of this repair mechanism was suggested by the work of John Cairns and his colleagues who discovered a repair protein in *Escherichia coli* with the required specifity for 0^6-methylguanine in DNA (Cairns, 1980). Extensive work on this inducible protein by Lindahl and his group (1988) revealed that it transfers the methyl group from 0^6-methylguanine in DNA to a sulfhydryl group of its own and thus inactivates itself. This so-called suicide protein was named 0^6-alkylguanine-DNA alkyltranferase and has been found to occur in all mammalian species investigated including the human (Pegg, 1990). Taken together, the above findings have led to the view, almost the belief, that the formation of 0^6-methylguanine in the DNA of target organs of animals treated with methylating carcinogens is a necessary, if not sufficient, cause of their carcinogenic action (Pegg and Singer, 1984). This proposal will be considered with reference to the restructured Koch's Postulates of Hall and Lemoine (1991) in the light of recent work on the molecular and cell biology of carcinogenesis.

The convincing demonstration of reaction of activated chemical carcinogens with DNA gave considerable support for the somatic mutation theory of cancer proposed much earlier in 1914 by Boveri (Boveri, 1914). The later demonstration by Bruce Ames that many carcinogens are also mutagens after suitable metabolic activation provided further evidence (Ames et al., 1973). It was not, however, until the discovery that the neoplastic phenotype could be induced in cultured cells by transfection of purified DNA from several human and chemically-induced animal tumors (Shih et al., 1979; Kontiris and Cooper,

1981) that the actual presence of mutant genes in tumors was finally shown. The current hypothesis is, therefore, that certain genes described as proto-oncogenes, which are present in normal cells, become mutated to form oncogenes which, in some way, cause the development of the malignant phenotype (Bishop, 1983). In other cases, the mutation occurs in so-called tumor suppressor genes or antioncogenes which, when deactivated by mutation, also result in neoplastic transformation (Weinberg, 1989). In some cases the mutation in the proto-oncogene or the suppressor gene is thought to arise from covalent interaction with the active form of a carcinogen.

Some of the most convincing evidence for this idea has come from work on tumors in experimental animals induced by N-nitroso compounds, in particular the rat mammary tumor induced by a single dose of N-methylnitrosourea (Thompson and Meeker, 1983), a compound which decomposes in the body after injection in less than 20 minutes (Swann, 1968). This model was used very effectively by Saraswati Sukumar and colleagues (1983) to demonstrate the presence of a mutated *ras* oncogene in the established tumors. The mutation was shown to be in the 12th codon of the gene, where the normal triplet GGA was changed to GAA. An identically changed triplet had been found earlier in human bladder tumors (Reddy et al., 1982). It was of particular interest that the type of mutation found, a G → A transition, is exactly that expected to result from interaction between methylnitrosourea and DNA (Singer and Kusmierek, 1982). It was therefore suggested that the tumors could have been initiated by methylation of this oncogene during the brief exposure to the pulse dose of the carcinogen (Sukumar et al., 1983). An alternative explanation, recognized by Sukumar and colleagues, would have been that the *ras* mutation occurred during the development stages of the tumors and was thus not related directly to the DNA reaction with the very short-lived carcinogen. This question was addressed by the same group (Kumar et al., 1990), who exposed rats to methylnitrosourea at birth and examined the mammary glands two weeks after carcinogen treatment. High resolution restriction fragment length polymorphism of polymerase chain reaction-amplified *ras* sequences showed the presence of both H-*ras* and K-*ras* as least two months before what was considered to be the onset of neoplasia. The *ras* oncogene appeared to remain latent within the mammary gland until exposure to estrogens, suggesting that normal physiological proliferative processes such as estrogen-induced mammary gland development may lead to neoplasia if the targeted cells contain latent *ras* oncogenes.

Support for the idea that activated *ras* oncogenes play a part in the induction of rat mammary carcinoma has come from work by Michael Gould and colleagues who have developed a model system for introducing genes into the mammary epithelial cells of intact rats using replication-defective retrovirus vector-mediated gene transfer (Wang et al., 1991a, b). In these experiments, biologically activated genetic vectors containing the v-Ha-*ras* or the *neu* oncogenes were infused into the central ducts of the mammary glands. The injected material thus came into direct contact with and penetrated the mammary epithelial cells. Spread of virus infection was prevented by the defect in virus replication and absence of helper virus. The recombinant retroviruses were identical apart from the presence of a 700-base-pair v-Ha-*ras* or the entire complementary DNA of the *neu* gene. The tumors induced by the *ras* oncogene resembled those induced by N-methylnitrosourea and other chemical carcinogens in both the kinetics of their development and widely varying histopathological spectrum. Most of those following *neu* oncogene transfer, however, were uniform in morphology and were classified as cribriform-comedocarcinomas. Significantly, the *neu* gene was more potent than the *ras*, under these conditions, by about two orders of magnitude, leading to the conclusion that the transformation of rat mammary cells by *ras* oncogenes requires additional genetic factors (Wang et al., 1991b).

The above work shows that transfection of the v-*ras* oncogene can induce mammary cancer *in vivo* in the intact animal. It does not, of course, give direct information on effects of a single mutation in the 12th codon of the gene, presumed to result from formation of 0^6-methylguanine by exposure to N-methylnitrosourea. This question has been addressed by Mitra and co-workers (1989) using a synthetic rat c-H-*ras* gene containing 0^6-methylguanine and by Kamiya and co-workers with the human gene, each incorporated into appropriate transfection vectors. In each case transfection of the modified genes induced cellular transformation (Mitra et al., 1989; Kamiya et al., 1991).

As mentioned before, 0^6-methylguanine residues in DNA are repaired by the repair protein 0^6-alkylguanine-DNA alkyltransferase (Pegg, 1990). Thus, induced alterations in the activity of this protein would be expected to influence toxic, mutagenic, and carcinogenic effects resulting from the formation of this DNA adduct (Pegg and Singer, 1984). The discovery that the transferase is strongly inactivated by 0^6-benzylguanine (Dolan et al., 1990) provided a means to investigate the role of 0^6-methylguanine DNA-adducts in cytotoxicity, mutagenesis, and carcinogenesis. Very recently, Lijinsky and colleagues (1994) have investigated the effects of treatment with 0^6-benzylguanine, which has relatively low toxicity, on carcinogenesis by methylnitrosourea or ethylnitrosourea in rats. The animals received the nitrosoureas over 10 weekly doses, and the test group were given 0^6-benzylguanine 2 hours before the carcinogens at doses sufficient to reduce the alkyltransferase activity to zero for at least 8 hours. This regime had no significant effect on the incidence of the principal induced tumors and minimal effect on the induction times (Lijinsky et al., 1994). The authors point out, however, that the reduction in alkyltransferase activity may not have lasted long enough to affect the repair of the adducted DNA sufficiently.

Stanton Gerson and colleagues have investigated the role of 0^6-methylguanine-DNA methyltransferase by a different approach using transgenic mice expressing high concentrations of the human alkyltransferase gene MGMT in the thymus. This organ is a major target for carcinogenesis by methylnitrosourea in the mouse, resulting in thymic lymphomas. The transgenic animals were effectively protected against the carcinogenic effects of methylnitrosourea in the thymus, but surprisingly, the incidence of splenic lymphomas was virtually unaffected by the presence of the transgene. The protective effect of the transgene against thymic lymphoma is supported by the reported equal frequency of radiation-induced lymphomas in MGMT[+] and nontransgenic littermates (Dumenco et al., 1993). Consistent results were obtained by Nakatsuru and colleagues (1993) using a transgenic mouse strain that had carried the *E. coli* alkyl-DNA transferase gene *ada*, attached to the Chinese hamster metallothionein 1 gene over many generations. The livers of these mice consistently showed 3 times the control transferase activity which could be increased to 6 to 8 times by injection of $ZnSO_4$. Significantly reduced liver tumor formation was observed in animals treated with dimethyl- or diethylnitrosamine, the differential effect not appearing at the higher dose levels (Nakatsuru, 1993).

It thus appears that there is substantial evidence that the methylating nitroso carcinogens act by formation of alkyl adducts on the 6-position of guanine in DNA, with less evidence for the ethylating compounds. The specificity of the mammalian repair protein for this position, unlike the bacterial protein (Pegg, 1990), underlines the probable major significance of this adduct for carcinogenesis.

It would be tempting to conclude that interaction with cellular DNA is the causative lesion in carcinogenesis by other N-nitroso compounds and, indeed, by the many other types of chemical carcinogen that have been shown to react covalently with DNA. However, the evidence for causation is considerably less than with the methylating nitroso compounds, and this applies to the complex N-nitrosamines and other carcinogens carrying N-nitroso groups. This apparent discrepancy has been emphasized by Lijinsky, who has studied the structure-activity relationships of a very large number of N-nitroso carcinogens. His views were expressed succinctly in 1988 as follows: "The conclusion of this survey is that measurements of alkylation of DNA by a carcinogenic N-nitroso compound or similar alkylating agent does not provide substantial information about the mechanism of tumor induction by these compounds ..., other factors are of much greater importance ... but what they are is unknown" (Lijinsky, 1988). Space does not permit adequate presentation of the data on which these views are based, but the reader is referred to the comprehensive monograph by Lijinsky (1992).

IV. GENOTOXIC AND NONGENOTOXIC CARCINOGENS

The concept of genotoxicity and its absence in nongenotoxic or epigenetic carcinogens derives largely from the recognition that some carcinogens, in their activated forms, bind covalently to DNA. The term genotoxic was proposed in 1971 at a meeting in Sweden on the Evaluation of Genetic Risks of

Environmental Chemicals (Brookes et al., 1973), but was not widely approved at first. It has gradually gained wider acceptance, largely through the efforts of Gary Williams and John Weisburger and is now part of the language of the subject. With the passage of time, the definition of genotoxicity has become blurred. The term has been used to include carcinogens that react chemically with DNA and/or induce mutations, or chromosomal aberrations, or have clastogenic activity. Recently Williams has recommended that "DNA reactive" should replace genotoxic and that carcinogens without DNA reactivity should be described as epigenetic (Williams, 1992). Space does not permit further discussion and attempts at clarification of these terms. Only a very brief comment will be made on their importance in risk assessment of carcinogens. Many human tumors are of clonal origin (Fialkow, 1976) as are tumors induced in experimental animals. For example, E.D. Williams and colleagues (1983) induced liver tumors with diethylnitrosamine followed by phenobarbital in female Sparse Fur (SpF) mice hererozygous for the x-linked enzyme ornithine carbamoyl transferase. Using a histochemical method capable of revealing the normal variant only, they found that the majority of the tumors showed only one or the other phenotype, implying an origin from single cells by clonal expansion (Nowell, 1976).

The clonal origin of some chemically induced tumors suggests they may have different dose-response relationships from the usual pharmacological responses which have a definite threshold. If, as much evidence discussed above suggests, some tumors do arise by clonal expansion from cells that have been mutated in oncogenes or suppressor genes by genotoxic chemicals, the dose required to produce the original mutation could be very small indeed since replication of the mutated cell would not necessarily be dependent on the presence of the original chemical. The cell proliferation required for the expansion might proceed autonomously, or as a result of other chemical stimuli (Butterworth and Slaga, 1991), for which a threshold could exist. Such chemicals would include nongenotoxic or epigenetic chemicals as defined by Williams (1992). The situation is not sufficiently clear for definite conclusions to be drawn, but further investigation in this area could be worthwhile.

V. PRACTICAL ASPECTS OF DNA ADDUCT FORMATION *IN VIVO*

A. IMPLICATIONS OF GENOTOXICITY FOR SAFETY EVALUATION. THE CASE OF TAMOXIFEN

There is no general agreement on the implications of genotoxicity, particularly covalent DNA binding, for the assessment of human hazard. This situation may be resolved in the foreseeable future when the outcome of exposure of large numbers of women to the drug tamoxifen over long periods becomes known. Tamoxifen is a nonsteroidal triphenylethylene derivative with a chemical structure similar to diethylstilbestrol, which has been used extensively in the treatment of early and advanced breast cancer. Women treated with tamoxifen were found to have a lower incidence of second and contralateral breast tumors, which led to the proposal that the drug should be used as a chemopreventive agent in women without apparent disease but with a high risk of developing one.

Unfortunately, animal tests indicated that tamoxifen was a strong liver carcinogen in rats (Williams et al., 1993) and two groups have shown concurrently that it forms covalent adducts with DNA in the livers of treated animals (Han and Liehr, 1992; White et al., 1992), both using the ^{32}P-postlabeling procedure. Furthermore, tamoxifen induced a significant increase in micronucleus formation in a dose-dependent manner in cultures of MCL-5 cells, a human cell line that expresses five different human cytochrome P450 isoenzymes, as well as epoxide hydrolase. Very recently the bound tamoxifen of one modified base was shown to be an epoxide-deoxyguanylic acid adduct (Phillips et al., 1994).

A clinical trial of tamoxifen for the prevention of human breast cancer was started in 1982. It has involved thousands of healthy young women at several hundred collaborating clinical centers in North America. The trial was suspended recently because audits showed signs of lax administration. A smaller trial is in progress in the United Kingdom. Concern has been expressed, however, regarding toxic and carcinogenic risks in human subjects, and an increased incidence of endometrial cancer has been reported (Gusberg, 1990). As pointed out by Han and Liehr (1992), it is possible that tamoxifen may generate DNA damage in the human uterus and that these covalent DNA modifications may initiate tumor development. These authors suggest that tamoxifen may be a poor choice for the chronic preventive treatment of breast cancer.

These criticisms derive added point from the recent comparison by Hard and colleagues (1993) of tamoxifen with a structurally related nonsteroidal triphenylethylene derivative, toremifene, which differs from tamoxifen only by the substitution of a chlorine atom for a hydrogen in an ethyl group attached to a carbon of the ethylenic double bond. Toremifene has estrogenic and antiestrogenic effects similar to tamoxifen and is undergoing clinical trial for treatment of breast cancer in Europe and the United States.

In marked contrast with tamoxifen, toremifene induced no liver tumors in rats exposed to dose levels comparable to those of tamoxifen, which caused liver cancer in all the treated animals. The expected DNA adducts of tamoxifen were demonstrated, whereas none could be found after toremifene treatment (Hard et al., 1993).

In light of the above accumulated evidence, the wisdom of exposing large populations of apparently healthy young women over long periods to pharmacologically active levels of tamoxifen has, perhaps not surprisingly, been questioned (Williams et al., 1992; Marshall, 1994).

B. DNA ADDUCTS AS MARKERS OF EXPOSURE
TO CARCINOGENS

As well as their intrinsic interest for the better understanding of molecular mechanisms of chemical carcinogenesis, DNA adducts have great practical value as markers of exposure to environmental carcinogens. As already indicated, the DNA adducts of a considerable number of known carcinogens have been identified and their chemical structures determined. Experiments with labeled carcinogens have clearly shown that the moieties adducted to DNA are derived from the administered agent. Their presence, therefore, indicates that the compound has penetrated within the cells of the organ or tissue studied and undergone metabolic activation if needed. These considerations have given rise to the idea of the biologically active dose, which reflects the amount of carcinogen that has interacted with cellular macromolecules at a target site or with an established surrogate (Perera and Santella, 1993). Clearly, to be of practical use, DNA from human subjects must be readily obtainable by noninvasive or minimally invasive procedures. Sampling is thus mainly limited to blood, placenta, and urine, with occasional use of liver or other tissues. Reliable methods are available for obtaining adequate amounts of DNA from blood leukocytes, and DNA bases with attached adduct can be obtained from urine. These adducted bases are derived from the DNA by processes of enzymic DNA repair or by simple chemical decomposition and, in those cases tested, they appear to be excreted unchanged in the urine (Craddock and Magee, 1967). Analytical methods have been progressively improved and are now sufficiently refined to permit the detection and estimation of minute quantities of adducts. The methods available include physicochemical and immunochemical procedures, as well as the technique for ^{32}P-postlabeling (Randerath et al., 1981). Physicochemical methods include high performance liquid chromatography with ultraviolet or fluorescence detection, atomic absorption spectrometry, synchronous scanning fluorescence spectrometry, electrochemical detection, and tandem mass spectrometry. The estimated lower limit of detection with these procedures ranges from 100,000 fmol/µg DNA, or 1×10^{-5} adducts per base, to 0.5 fmol/µg DNA or 1×10^{-9} adducts/base for tandem MS. Various forms of immunochemical procedures range in sensitivity from 5 to 1 fmol/µg DNA (Lohman, 1988). (See also Chapter 6 by P.B. Farmer in this book.)

All of the above methods require prior knowledge of the identity of the DNA adduct. ^{32}P-Postlabeling detects the presence of nucleotides, presumably with attached adducts, not found in apparently normal DNA. The procedure was devised by Kurt Randerath and associates and involves the enzymic incorporation of ^{32}P into the constituent nucleotides of DNA after its isolation and enzymic digestion, followed by thin-layer or other chromatographic separation (Randerath et al., 1981). This method requires only 0.01 µg DNA but current purification methods usually require at least 1 µg DNA (Lohman, 1988).

A substantial amount of published work has now accumulated on DNA adducts found in human subjects (Poirier and Weston, 1991), much related to aromatic compounds. Only very few examples of this will be given, again reflecting the interests of the writer. For obvious reasons, by far the greater part of the published data refers to DNA from human blood leukocytes; however, there have been several reports of measurements on other organs or tissues.

An unusual example of carcinogen-adduct formation was reported by Herron and Shank (1980), who had the opportunity to examine the liver of a human subject who had died of suspected acute poisoning by dimethylnitrosamine. Liver DNA from the victim shortly after death showed the presence of 7-methyl and O^6-methylguanine, readily detectable by HPLC with fluorescence detection. This was obviously a unique case, and others have involved much smaller amounts of adducts. For example, Umbenhauer and colleagues detected the presence of O^6-methylanamine in DNA samples from esophageal tissues of human subjects living in the Linxian area of China, where the incidence of esophageal cancer is particularly high. Samples from Chinese populations with lower incidence of esophageal cancer had lower levels of O^6-methylguanine (Umbenhauer et al., 1985). Further examples of adduct formation in human tissues other than blood and placenta include the demonstration of 4-aminobiphenyl adducts in surgical samples of histologically normal lung and urinary bladder from smokers (Bartsch et al., 1988), benzo(a)pyrene diolepoxide adducts in lung (Bartsch et al., 1988), and aflatoxin-B_1 adducts in the livers of Chinese populations exposed to aflatoxin in their diet (Zhang et al., 1992).

Additional human data are available relating to the more readily accessible placental (Foiles et al., 1988) and buccal mucosal tissues, and very much more from human blood leukocyte DNA. This has been extensively discussed in several recent publications (Zhang et al., 1992; Harris, 1991; Schulte and Perera, 1993).

A major question arises concerning the relationship between levels of DNA adducts in blood leukocytes and those in other organs and tissues. These levels must be dependent on capacity to activate the carcinogen and to repair the adduct. Therefore, those found in the blood may give unreliable indications of levels in the organs in which tumors appear. Since the amounts of the P450 enzymes vary extensively from organ to organ, and may be very low in blood cells, the further question arises how adducts are formed in blood. An approach to answering these questions has been made recently by Bianchini and Wild (1994), who compared 7-methylguanine formation in white blood cells, liver, and target organs in rats treated with methylating carcinogens. They found that the ratio between 7-methylguanine formation in target organs and white blood cells varied considerably with the different carcinogens, whereas the ratio of 7-methylguanine in the liver DNA to that in the blood cells varied much less. They concluded that the presence of 7-methylguanine in white blood cell DNA is indicative of adduct formation in internal organs, particularly in the liver, and concluded that active methylating species may be formed in the liver and then transferred to the blood cells. Similar *in vivo* studies appear not to have been done with other types of carcinogens, but it seems reasonable to believe that the same conclusions may apply to them.

There is thus a substantial database on quantitative measurements of DNA adducts in blood and other organs of humans exposed to a variety of environmental carcinogens. The interpretation and implications of this information are rather uncertain, and Kaldor and Day, writing in 1988 (Bartsch et al., 1988) commented, "This field is nonetheless very young and has so far yielded few studies that would stand up to vigorous examination by epidemiological criteria." Subsequent research seems not to have invalidated this view, and future work might be better aimed at meeting such epidemiological criteria rather than the accumulation of more and more data on a less adequately planned basis.

C. DNA ADDUCT DOSIMETRY AND RISK ASSESSMENT

The examples discussed above of measurement of DNA adduct levels in human leukocytes and other tissues provide objective and quantitative evidence of exposure to environmental carcinogens, but only limited information on dose-response relationships, since the concentrations of the carcinogens in the environment are known only imprecisely, if at all. In one situation, however, the human exposure levels are known with considerable precision. Some chemicals used in the chemotherapy of human cancer are known to be carcinogenic in animals and to induce so-called second tumors in patients. Such chemicals include the platinum-based drugs and various alkylating agents including the nitrogen mustards, various chloroalkylnitroso-ureas, and the hydrazino derivatives procarbazine and dacarbazine. Procarbazine, or N-isopropyl-α-(2-methylhydrazino)-ρ-toluamide), has been used extensively for this purpose, at first alone, but now almost always as part of the drug combination known as MOPP (mechloroethamine, vincristine sulfate, prednisone, and procarbazine). Procarbazine is carcinogenic in mice, rats, and monkeys, inducing hemo-lymphoid, mammary, lung, and nervous system tumors (Bianchini and Wild, 1994). Human subjects successfully treated with MOPP for Hodgkin's lymphoma have a significantly increased risk of developing nonlymphocytic leukemia later (Levine and Bloomfield, 1992). It cannot be said with certainty that procarbazine is responsible, solely or in part, for the causation of human neoplasia by MOPP, but it is the only component of the mixture that can act as a methylating agent. Procarbazine is metabolized by rather complex pathways to yield methylazoxy-procarbazine as a major circulating metabolite which yields the methyldiazonium ion, $CH_3N_2^+$, in the tissues and acts as an S_N1 type methylating agent in the same way as N-methylnitrosourea and dimethylnitrosamine. Unlike these simple methylating agents, procarbazine also yields another diazonium ion, $RCH_2N_2^+$, where R represents the non-methylhydrazino moiety of the molecule. The biological effects of these intermediates are not known (Foiles et al., 1988).

Louise Fong, David Jensen and the writer (1990) compared the dosimetry of O^6-methyl-guanine and 7-methylguanine in female rats given single doses of procarbazine or N-methylnitrosourea that, under conditions causing depletion of alkyltransferase, induced mammary tumors. There was reasonable correlation between the amounts of 0^6-methylguanine produced in the mammary gland and the tumor yield. The same group studied the effects of repeated doses of procarbazine on various rat tissues and found a good correlation between 0^6-methylguanine and alkyltransferase depletion in mammary gland, lymph nodes, and thymus, the targets for tumor induction (Fong et al., 1992), supporting the view that accumulation of 0^6-methylguanine plays an important part in tumor induction by procarbazine. Similar results have been very recently reported by Kyrtopoulos and colleagues (Souliotis et al., 1994; Valavanis et al., 1994) where rats were given daily oral doses of procarbazine at levels similar to those received by human patients undergoing chemotherapy for Hodgkin's lymphoma. These authors point out that measurement of accumulation and repair of 0^6-methylguanine in the DNA of experimental animals, as well as in human subjects, treated with similar dose regimes of procarbazine "offers a unique opportunity to compare human and rodent susceptibilities to the formation of this important precarcinogenic adduct" (Souliotis et al., 1994). The same group (Valavanis et al., 1994) compared the accumulation of 0^6-methylguanine in 21 Hodgkin's lymphoma patients receiving MOPP therapy containing 150 mg procarbazine at one dose per day with rats given a range of doses of the drug once a day spanning this level. They found that, under conditions of no alkyltransferase depletion, 0^6-methylguanine accumulated in human blood leukocyte DNA at a rate about twofold lower than in rats suggesting that, to the extent that 0^6-methylguanine contributes to its carcinogenic activity, the susceptibility of human subjects to procarbazine is likely to be comparable to that of the rat. Since the same methyldiazonium ion intermediate is formed from dimethylnitrosamine and other methylating nitroso compounds, it may be reasonable to use levels of DNA methylation in humans receiving carcinogenic dose regimes of procarbazine to get a better quantitative insight into implications and possible risks of minute amounts of 0^6-methylguanine found in human populations exposed to N-nitroso compounds in the environment, for example, Chinese people living in areas of high risk for esophageal cancer (Umbenhauer et al., 1985) and female smokers (Foiles et al., 1988).

VI. CONCLUSIONS

The fact of covalent interaction of some chemical carcinogens with DNA, the recognition that many of them are mutagenic in their activated forms, and the discovery of mutated oncogenes in many human and experimentally induced animal tumors give strong support for the somatic mutation theory of the origin of cancer, proposed by Theodore Boveri in 1914.

Some carcinogens, however, are not mutagenic and do not react detectably, or to any great extent, with DNA. These differences have given rise to the idea of two broad categories of carcinogens, described as genotoxic or DNA reactive and nongenotoxic or epigenetic. This distinction may have important implications for the possible existence of threshold, and thus acceptable, levels of environmental carcinogens.

Apart from the mechanistic implications of carcinogen-DNA interactions, their detection in human DNA may have great value as an indication of exposure to environmental carcinogens in the workplace or elsewhere.

In the special case of human exposure to chemical carcinogens in known amounts in the form of cancer chemotherapeutic drugs, important conclusions may be drawn for risk assessment. Such patients may develop second tumors attributable to the treatment and comparisons of adduct levels in their DNA with those in experimental animals receiving carcinogenic dose regimes of the same compound may allow improved extrapolation of animal data to man.

While the exact significance of interaction of environmental chemicals with human DNA may not be known, it would, perhaps, be prudent to bear in mind the quotation, attributed to Mortimer Mendelson by Kari Hemminki (1993) as follows: "I wouldn't like to have my DNA messed up."

REFERENCES

Ames, B. N., Durston, W. E., Yamasaki, E., and Lee, F. D., Carcinogens are mutagens: a simple test system combining liver homogenates for activation and bacteria for detection, *Proc. Natl. Acad. Sci. U.S.A.*, 70, 2281, 1973.

Bartsch, H., Hemminki, K., and O'Neill, I. K., Eds., *DNA Damaging Agents in Humans: Applications in Cancer, Epidemiology and Prevention*, International Agency for Research on Cancer, Publication 89, Lyons, 1988.

Bianchini, F. and Wild, C. P., Comparison of 7-medG formation in white blood cells, liver and target organs in rats treated with methylating carcinogens, *Carcinogenesis*, 15, 1137, 1994.

Bishop, J. M., Cellular oncogenes and retroviruses, *Ann. Rev. Biochem.*, 52, 301, 1983.

Boveri, T., *Zur Frage der Entstehung Maligner Tumoren*, Gustave Fischer Verlag, Jena, 1914.

Boyland, E., Different types of carcinogens and their possible modes of action, *Cancer Res.*, 12, 77, 1952.

Brookes, P., Druckrey, H., Ehrenberg, L., Lagerlof, B., Litwin, J., and Williams, G. M., The relation of cancer induction and genetic damage, in *Evaluation of Genetic Risks of Environmental Chemicals*, Ramel, C., Ed., Ambio Special Report No. 3, 1973, 15.

Brookes, P. and Lawley, P. D., The reaction of mustard gas with nucleic acids *in vitro* and *in vivo*, *Biochem. J.*, 77, 478, 1960.

Brookes, P. and Lawley, P. D., Evidence for the binding of polynuclear aromatic hydrocarbons to the nucleic acids of mouse skin: relation between carcinogenic power of hydrocarbons and their binding to deoxyriboneucleic acid, *Nature (Lond.)*, 202, 781, 1964.

Butterworth, B. E. and Slaga, T. J., Eds., *Chemically Induced Cell Proliferation: Implications for Risk Assessment*, Wiley-Liss, New York, 1991.

Cairns, J., Efficiency of the adaptive response of *Escherichia coli* to alkylating agents, *Nature (Lond.)*, 286, 176, 1980.

Craddock, V. M. and Magee, P. N., Effect of administration of the carcinogen dimethylnitrosamine on urinary 7-methylguanine, *Biochem. J.*, 104, 435, 1967.

Dolan, M. E., Moschel, P. C., and Pegg, A. E., Depletion of mammalian O^6-alkylguanine-DNA alkyltransferase activity by O^6-benzylguanine provides a means to evaluate the role of this protein against carcinogenic and therapeutic alkylating agents, *Proc. Natl. Acad. Sci. U.S.A.*, 87, 5368, 1990.

Druckrey, H., Preussmann, R., Ivankovic, S., and Schmähl, D., Organotrope carcinogene Wirkungen bei, 65 verschiedenen N-Nitroso Verbingdungen an BD ratten, *Z. Krebsforsch*, 69, 103, 1967.

Dumenco, L. L., Allay, E., Norton, K., and Gerson, S. L., The prevention of thymic lymphomas in transgenic mice by human O^6-alkylguanine-DNA alkyltransferase, *Science,* 259, 219, 1993.

Farber, E. and Magee, P. N., The probable alkylation of liver ribonucleic acid by the hepatic carcinogens dimethylnitrosamine and ethionine, *Biochem. J.*, 76, 58P, 1960.

Fialkow, P. J., Clonal origin of human tumors, *Biochim. Biophys. Acta*, 458, 283, 1976.

Foiles, P. G., Miglietta, L. M., Akerkar, S. A., Everson, R. B., and Hecht, S. S., Detection of O^6-methyldeoxyguanosine in human placental DNA, *Cancer Res.*, 48, 4148, 1988.

Fong, L. Y. Y., Bevill, R. F., Thurson, J. C., and Magee, P. N., DNA adduct dosimetry and DNA repair in rats and pigs given repeated doses of procarbazine under conditions of carcinogenicity and human cancer chemotherapy respectively, *Carcinogenesis*, 13, 2153, 1992.

Fong, L. Y. Y., Jensen, D. E., and Magee, P. N., DNA methyl-adduct dosimetry and O^6-alkylguanine-DNA alkyl transferase activity determinations in rat mammary carcinogenesis by procarbazine and N-methylnitrosourea, *Carcinogenesis*, 11, 411, 1990.

Goth, R. and Rajewsky, M. F., Persistance of 06-ethylguanine in rat brain DNA: correlation with nervous system specific carcinogenesis by ethylnitrosourea, *Proc. Natl. Acad. Sci. U.S.A.,* 71, 639, 1974.

Gusberg, S. B., Tamoxifen for breast cancer: associated endometrial cancer, *Cancer (Phila.)*, 65, 1463, 1990.

Hall, P. A. and Lemoine, N. R., Koch's postulates revisited, *J. Pathol.*, 164, 283, 1991.

Han, X. and Liehr, J. G., Induction of covalent DNA adducts in rodents by tamoxifen, *Cancer Res.*, 52, 1360, 1992.

Hard, G. C., Iatropoulos, M. J., Jordan, K., Radi, L., Kaltenberg, O. P., Imondi, A. R., and Williams, G. M., Major difference in the hepatocarcinogenicity and DNA adduct forming ability between toremifene and tamoxifen in female Cri: CD (BR) rats, *Cancer Res.*, 53, 4534, 1993.

Harris, C. C., Chemical and physical carcinogenesis: advances and perspectives for the 1990s, *Cancer Res.,* 51, 5023s, 1991.

Heidelberger, C., Studies on the molecular mechanism of hydrocarbon carcinogenesis, *J. Cell Comp. Physiol.* (Suppl. 1), 64, 129, 1964.

Hemminki, K., DNA adducts, mutations and cancer, *Carcinogenesis*, 14, 2007, 1993.

Herron, D. C. and Shank, R. C., Methylated purines in human liver DNA after probable dimethylnitrosamine poisoning, *Cancer Res.*, 40, 3116, 1980.

Kamiya, H., Miura, K., Ohtomo, N., Nishimura, S., and Ohtsuka, E., Transforming activity of a synthetic c-Ha-ras gene containing O^6-methylguanine in codon 12, *Jpn. J. Cancer Res.*, 82, 997, 1991.

Kontiris, T. G. and Cooper, G. M., Transforming activity of human tumor DNAs, *Proc. Natl. Acad. Sci. U.S.A.*, 78, 1181, 1981.

Kumar, R., Sukumar, S., and Barbacid, M., Activation of ras oncogenes preceding the onset of neoplasia, *Science*, 248, 1101, 1990.

Levine, E. G. and Bloomfield, C. D., Leukemias and myelodysplastic syndromes secondary to drug, radiation and environmental exposure, *Seminars Oncol.*, 19, 47, 1992.

Lijinsky, W., Nucleic acid alkylation by N-nitroso compounds related to organ-specific carcinogenesis, in *Chemical Carcinogens, Activation, Mechanisms, and Reactivity,* Politzer, P. and Roberts, L., Eds., Elsevier Science Publishers, Amsterdam, 1988, 242.

Lijinsky, W., *Chemistry and Biology of N-Nitroso Compounds*, Cambridge University Press, Cambridge, 1992, 464.

Lijinsky, W., Pegg, A. E., Anver, M. R., and Moschel, R. C., Effects of inhibition of O[6]-alkylguanine-DNA alkyltransferase in rats on carcinogenesis by methylnitrosourea and ethylnitrosourea, *Jpn. J. Cancer Res.*, 85, 226, 1994.

Lindahl, T., Sedgwick, B., Sekiguchi, M., and Nakabeppu, Y., Regulation and expression of the adaptive response to alkylating agents, *Ann. Rev. Biochem.*, 57, 133, 1988.

Lohman, P. H. M., Summary: adducts, in *DNA Damaging Agents in Humans*, Bartsch, H., Hemminki, K., and O'Neill, I. K., Eds., International Agency for Research on Cancer, Publication 89, Lyons, 1988.

Loveless, A., Possible relevance of 0-6 alkylation of deoxyguanosine to the mutagenicity and carcinogenicity of nitrosamines and nitrosamides, *Nature (Lond.)*, 223, 206, 1969.

Magee, P. N., The experimental basis for the role of nitroso compounds in human cancer, *Cancer Surveys*, 8, 207, 1989.

Magee, P. N. and Barnes, J. M., The production of maligant primary hepatic tumours in the rat by feeding dimethylnitrosamine, *Brit. J. Cancer*, 10, 114, 1956.

Magee, P. N. and Farber, E., Toxic liver injury and carcinogenesis. Methylation of rat liver nucleic acids by dimethylnitrosamine *in vivo*, *Biochem. J.*, 83, 114, 1962.

Margison, G. P. and Kleihues, P., Chemical carcinogenesis in the nervous system. Preferential accumulation of O[6]-methylguanine in rat brain deoxyribonucleic acid during repetitive administration of N-methyl-N-nitrosourea, *Biochem. J.*, 148, 521, 1975.

Marshall, E., Tamoxifen: hanging in the balance, *Science*, 264, 1524, 1994.

Miller, E. C. and Miller, J. A., The presence and significance of bound aminoazodyes in the livers of rats fed *p*-dimethylaminoazobenzene, *Cancer Res.*, 7, 468, 1947.

Miller, E. C. and Miller, J. A., Biochemistry of carcinogenesis, *Ann. Rev. Biochem.*, 28, 291, 1959.

Miller, E. C. and Miller, J. A., Mechanisms of chemical carcinogenesis: nature of proximate carcinogens and interactions with macromolecules, *Pharmacol. Revs.*, 18, 805, 1966.

Mitra, G., Pauly, G. T. Kumar, R., Pei, G. K., Hughes, S. H., Moschel, R. C., and Barbacid, M., Molecular analysis of O[6]-substituted guanine-induced mutagenesis in ras oncogenes, *Proc. Natl. Acad. Sci. U.S.A.*, 86, 8650, 1989.

Nakatsuru, Y., Matsukuma, S., Nemoto, N., Sugano, H., Sekiguchi, M., and Ishikawa, T., O[6]-Methylguanine-DNA methyltransferase protects against nitrosamine-induced hepatocarcinogenesis, *Proc. Natl. Acad. Sci. U.S.A.*, 90, 6468, 1993.

Nowell, P. C., The clonal evolution of tumor cell populations, *Science (Washington DC)*, 194, 23, 1976.

Pegg, A. E., Mammalian 0[6]-alkylguanine-DNA alkyltransferase: regulation and importance in response to alkylating carcinogenic and therapeutic agents, *Cancer Res.*, 50, 6119, 1990.

Pegg, A. E. and Singer, B., Is O[6]-alkylguanine necessary for initiation of carcinogenesis by alkylating agents?, *Cancer Invest.*, 2, 221, 1984.

Perera, F. P. and Santella, R., Carcinogenesis, in *Epidemiology, Principles and Practices*, Schulte, P. A. and Perera, F. P., Eds., Academic Press, San Diego, 1993, 277.

Phillips, D. H., Hewer, A., White, I. N. H., and Farmer, P. B., Co-chromatography of a tamoxifen epoxide-deoxyguanylic acid adduct formed in the livers of tamoxifen-treated rats, *Carcinogenesis*, 15, 793, 1994.

Pitot, H. C. and Heidelberger, C., Metabolic regulatory circuits and carcinogenesis, *Cancer Res.*, 23, 1694, 1963.

Poirier, M. C. and Weston, A., DNA adduct determination in humans, *Prog. Clin. Biol. Res.*, 372, 205, 1991.

Preussmann, R. and Stewart, B. W., N-Nitrosocarcinogens, in *Chemical Carcinogens*, Searle, C. E., Ed., American Chemical Society Monograph, No. 182, Washington D. C., 1984, 182.

Randerath, K., Reddy, M. V., and Gupta, R. C., [^{32}P]-Labeling test for DNA damage, *Proc. Natl. Acad. Sci. U.S.A.*, 78, 6126, 1981.

Reddy, E. P., Reynolds, R. K., Santos, E., and Barbacid, M., A point mutation is responsible for the acquisition of transforming properties by the T24 human bladder cancer oncogene, *Nature (Lond.)*, 300, 149, 1982.

Rosenkranz, H. S. and Klopman, G., Use of a composite polyfunctional model electrophile as a probe to analyse the performance of an artificial intelligence structure-activity method, *Mutation Res.*, 232, 249, 1990.

Schulte, P. A. and Perera, F. P., Eds., *Molecular Epidemiology: Principles and Practices*, Academic Press, San Diego, 1993, 588.

Shih, C. Shilo, B. Z., Goldfarb, M. P., Dannenberg, A., and Weinberg, R. A., Passage of phenotypes of chemically transformed cells via transfection of DNA and chromatin, *Proc. Natl. Acad. Sci. U.S.A.*, 76, 5714, 1979.

Sims, P., Grover, P. I., Swaisland, A., Pal, K. and Hewer, A., Metabolic activation of benzo[a]pyrene proceeds by a diol-epoxide, *Nature (Lond.)*, 252, 326, 1974.

Singer, B., All oxygens in nucleic acids react with carcinogenic ethylating agents, *Nature (Lond.)*, 264, 333, 1976.

Singer, B. and Kusmierek, J. T., Chemical mutagenesis, *Ann. Rev. Biochem.*, 52, 655, 1982.

Souliotis, V. L., Valavanis, C. Boussiotis, V. A., Pangalis, G. A., and Kyrtopoulos, S. A., Comparative dosimetry of O^6-methylguanine in humans and rodents treated with procarbazine, *Carcinogenesis*, 15, 1675, 1994.

Sukumar, S., Notario, V., Martin-Zanca, D., and Barbacid, M., Induction of mammary carcinomas in rats by nitroso-methylurea involves malignant activation of H-ras-1 by single point mutations, *Nature (Lond.)*, 306, 658, 1983.

Swann, P. F., The rate of breakdown of methyl methanesulphonate, dimethyl sulphate and N-methyl-N-nitrosourea in the rat, *Biochem. J.*, 110, 49, 1968.

Swann, P. F. and Magee, P. N., Nitrosamine induced carcinogenesis: the alkylation of nucleic acids of the rat by N-methyl-N-nitrosourea, dimethylnitrosamine, dimethylsulphate and methyl methanesulphonate, *Biochem. J.*, 110, 39, 1968.

Thompson, H. J. and Meeker, L. D., Induction of mammary gland carcinomas by the subcutaneous injection of 1-methyl-1-nitrosourea, *Cancer Res.*, 43, 1628, 1983.

Umbenhauer, D., Wild, C. P., Montesano, R., Saffhill, R., Boyle, J. M., Huh, N., Kirsten, U., Thomale, J., Rajewsky, M. F., and Lu, S. H., O^6-methyldeoxy-guanosine in oesophageal DNA among individuals at high risk of oesophageal cancer, *Int. J. Cancer*, 36, 661, 1985.

Valavanis, C., Souliotis, V. L., and Kyrtopoulos, S. A., Differential effects of procarbazine and methylnitrosourea on the accumulation of O^6-methylguanine and the depletion and recovery of O^6-alkylguanine-DNA alkyltransferase in rat tissues, *Carcinogenesis*, 15, 1681, 1994.

Wang, B., Kennan, W. S., Yasukawa-Barnes, J., Lindstrom, M. J., and Gould, M. J., Carcinoma induction following direct *in situ* transfer of v-Ha-ras into rat mammary epithelial cells using replication-defective retrovirus vectors, *Cancer Res.*, 51, 2642, 1991a.

Wang, B., Kennan, W. S., Yasukawa-Barnes, J., Lindstrom, M. J., and Gould, M. N., Frequent induction of mammary carcinomas following neu oncogene transfer into *in situ* mammary epithelial cells of susceptible and resistant rat strains, *Cancer Res.*, 51, 5649, 1991b.

Weinberg, R. A., Oncogenes, antioncogenes and the molecular bases of multistep carcinogenesis, *Cancer Res.*, 49, 3713, 1989.

Wheeler, G. P. and Skipper, H. E., Studies with mustards III. *In vivo* fixation of $C^{14}H_3$ in nucleic acid fractions of animal tissues, *Arch. Biochem. Biophys.*, 72, 465, 1957.

White, I. N. H., De Matteis, F., Davies, A., Smith, L. L., Crofton-Sleigh, C., Venitt, S., Hewer, A., and Phillips, D. H., Genotoxic potential of tamoxifen and analogues in female Fischer F344/n rats, DBA/2 and C57BL/6 mice and human MCL-5 cells, *Carcinogenesis*, 13, 2197, 1992.

Williams, E. D., Wareham, K. A. and Howell, S., Direct evidence for the single cell origin of mouse liver cell tumours, *Brit. J. Cancer*, 47, 723, 1983.

Williams, G. M., DNA reactive and epigenetic carcinogens, *Exp. Toxic. Pathol.*, 44, 457, 1992.

Williams, G. M., Iatropoulos, M. J., Djordjevic, M. V., and Kaltenberg, O. P., The triphenylethylene drug tamoxifen is a strong liver carcinogen in the rat, *Carcinogenesis (Lond.)*, 14, 315, 1993.

Williams, G. M., Iatropoulos, M. J., and Hard, G. C., Long-prophylactic use of tamoxifen: is it safe?, *Eur. J. Cancer Prevention*, 1, 386, 1992.

Zhang, Y. J., Chen, C. J., Lee, C. S., Haghighi, B., Yang, G. Y., Wang, L. W., Feitelson, M., and Santella, R., Aflatoxin B_1-DNA adducts and hepatitis B virus antigens in hepatocellular carcinoma and non-tumorous lung tissue, *Carcinogenesis*, 13, 2247, 1992.

Part III
Genetic and Cellular
Aspects in Toxicology

Chapter 8

THE STUDY OF GENES AND GENE EXPRESSION BY *IN SITU* HYBRIDIZATION: A POTENTIALLY USEFUL APPROACH IN TOXICOLOGY

Ian Lauder

CONTENTS

I. Background 147

II. DNA Preservation in Archived Tissues 148

III.. *In Situ* Hybridization 150
    A. Sensitivity 152
    B. Probe Preparation and Labeling 153
    C. Alternative Haptens 154
    D. Nonradioactive Labeling and Detection 154

IV. Immunoglobulin Light Chain mRNA as a Model NISH System 155

V. Choice of Appropriate Controls 156

VI. Applications of NISH Technology 156
    A. Cell Proliferation and Histone mRNA 156
    B. Detection of Apoptosis 158
    C. Other Techniques 159

VII. Conclusion 161

References 161

I. BACKGROUND

In many current toxicological investigations, the emphasis has become focused on mechanisms of toxicity and in particular on the molecular basis of such toxicity in response to an ever-increasing number of external agents. Traditional toxicology has involved assessment of the direct toxic effects on animal tissues at both the biochemical and morphological levels. Much valuable information has been gained by such studies, but doubts will always exist

0-8493-9229-2/95/$0.00+$.50

147

as to the extent to which toxicological investigations in experimental animals can be applied directly to humans. Molecular pathology is a new branch of pathology which attempts to determine the molecular basis of disease processes. It is not a new discipline in its own right but rather an assimilation of techniques and methodology taken from modern molecular biology and applied in many instances directly to human tissues.

A considerable limiting factor in all studies of human disease is the ethical restrictions which rightly limit experimentation on humans. Pathology departments are uniquely placed for investigations of human tissues, as they are the custodians of a vast archive of human tissues. Each year some 2.5 million tissue samples from patients are processed in the United Kingdom according to the Royal College of Pathologists Manpower Survey, 1991. Virtually all departments retain tissue samples for a minimum of 10 years so that the total U.K. archive is at least 25 million samples. Allowing for the extreme reluctance of many pathology departments to part with their paraffin blocks, a true estimate of the nation's total archive is probably in excess of 50 million samples in all. These samples represent a unique record of the interaction of the environment with human patients, which in some departments goes back for a hundred years or more.

Much could be gained in terms of understanding the pathobiology of human toxicity by approaches which allowed genes and gene expression to be studied in archived human tissue samples. The molecular pathology research group was established at Leicester in 1984. While not directed specifically at toxicological mechanisms, much of the work of the research group is directly relevant to a modern mechanistic approach to human toxicity. It was appreciated immediately that the key to investigations on archived tissue lay in the degree of preservation of nucleic acids in archived material. Our initial investigations were designed to look at DNA and RNA preservation in archived samples.

II. DNA PRESERVATION IN ARCHIVED TISSUES

Using fresh human tonsil as a control tissue, experiments were performed to study the effects of fixation and tissue processing on DNA extracted from paraffin blocks of tissue. Tissue was fixed for periods varying from 0 to 72 hours in a variety of fixatives including formal-saline, buffered formalin, Carnoy's fixative, and Bouin's fixative. DNA was extracted by a standard phenol/chloroform procedure and, after restriction endonuclease treatment, electrophoresis, and Southern blotting, DNA preservation was assessed with a variety of ^{32}P-labeled probes (Warford et al., 1988).

A typical result is shown in Figure 8.1. The experiments revealed that in samples fixed for less than 12 hours, DNA can be extracted which is certainly usable for Southern blot analysis, but is always considerably inferior to DNA extracted from fresh tissue. In DNA extracted from lymphoid tissues, it was shown that rearrangements of immunoglobulin and T-cell receptor genes can sometimes be demonstrated (Figure 8.2). The results, however, indicated that

C   F
12345 12345

Key
C   Carnoy fixation
F   Formal saline fixation
1   24 hr fixation, paraffin
2   96 hr fixation, paraffin
3   168 hr fixation, paraffin
4   24 hr fixation, unprocessed
5   168 hr fixation, unprocessed

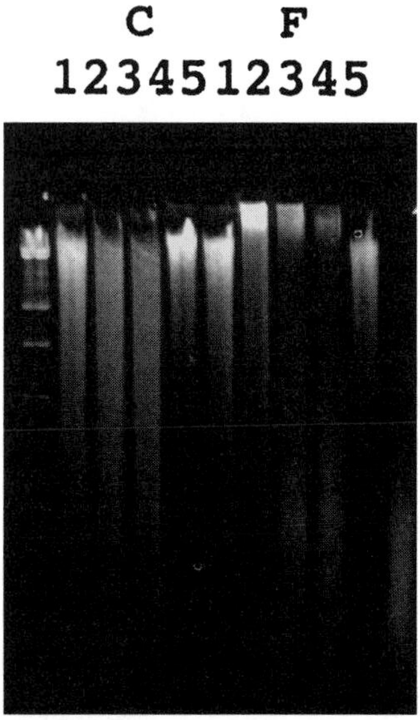

FIGURE 8.1. Electrophoresis of DNA extracted after a variety of fixation times using normal human tonsil immersed in Carnoy's fixative (C 1-5) and formal saline fixative (F 1-5). After electrophoresis and treatment with ethidium bromide, the tracks were visualized under UV light. The experiment showed that preservation of DNA was much better in Carnoy's fixative with a reasonable yield of high molecular weight DNA even after 168-hour fixation. After formalin fixation only the 24-hour fixed sample yielded usable high molecular weight DNA, with the 48-hour and 168-hour samples showing only low molecular weight fragments. DNA degradation also seems to be enhanced by tissue processing, as yields of high molecular weight DNA are higher in the unprocessed samples for both fixatives. (Figure kindly provided by A. Warford.)

beyond 12 hours fixation the DNA became increasingly degraded into low molecular weight fragments. Since most of the archival samples have been fixed for periods of time ranging from 12 to 24 hours, it was apparent that direct analysis of DNA from paraffin blocks of tissue would not be rewarding. We were, however, able to show that with some sources of DNA, such as viruses, acceptable results could be obtained, and these experiments did enable us to confirm that human papilloma virus DNA could still be recovered from paraffin blocks archived for more than 20 years. We were subsequently able to show that Epstein-Barr virus DNA extracted and blotted from cases of Hodgkin's Disease (Libetta et al., 1990) was also identifiable by a PCR-based strategy in the same cases (Brocksmith et al., 1991). It has of course subsequently been shown by numerous groups that despite the fragmentation which accompanies fixation, the DNA remains suitable for PCR analysis (Koopmans et al., 1993; Turner et al., 1993; Volkenandt et al., 1992). This paper will not however look at PCR, which could form the basis of a whole chapter on its toxicological applications. As such it is beyond the scope of this relatively short communication and suitable reviews are readily available (Kawasaki, 1992; Young, 1992).

III. *IN SITU* HYBRIDIZATION
(ISH; NISH IN NONRADIOACTIVE SYSTEMS)

Having established the unsuitability of extracted DNA from fixed tissues for genetic analysis by conventional Southern blot analysis, our attention next turned to studies of genes and gene expression in tissue sections (Warford and Lauder, 1991). The identification of nucleic acid sequences in tissue sections using labeled probes is not a new technique, having first been described some 25 years ago (Gall and Pardue, 1969; John et al., 1969). Much of this early work utilized cloned, double-stranded DNA probes which were usually labeled with radioactive isotopes such as ^{32}P or ^{35}S. A variety of probe types is possible (Table 8.1), and each has advantages and disadvantages. In our molecular pathology laboratory, we were particularly interested in the possible application of nonradioactively labeled (NISH) oligonucleotide probes to studies of

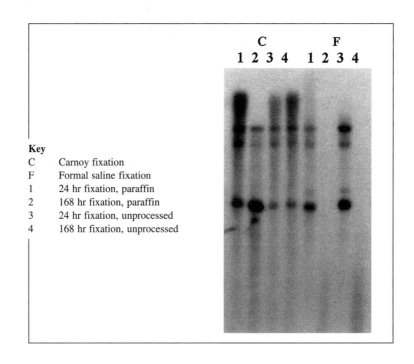

Key

C	Carnoy fixation
F	Formal saline fixation
1	24 hr fixation, paraffin
2	168 hr fixation, paraffin
3	24 hr fixation, unprocessed
4	168 hr fixation, unprocessed

FIGURE 8.2.   Hybridization of DNA extracted from fixed tissue. EcoR1 restricted DNA hybridized with Tβ probe. In this experiment DNA was extracted from tonsils after fixation in Carnoy's fixative or formal saline. After restriction enzyme treatment and electrophoresis, Southern blotting was performed and the filters were probed for the Tβ component of the T-cell receptor protein. As in Figure 1, preservation is again better in Carnoy's fixative than in formalin, and clear bands after autoradiography can be seen in all of the Carnoy's tracks (1-4). In formalin-fixed tissue, only the 24-hour fixed tissue revealed definite bands. Tissue processing again results in further apparent degradation. (Figure kindly provided by A. Warford.)

TABLE 8.1
Probe Types and Advantages/Disadvantages

Probe type	Advantages	Disadvantages
Double-stranded DNA	Long sequences possible Many bases can be labeled	Requires cloning Poor hybridization efficiency Poor probe penetration because of probe size
Single-stranded RNA	Long strands possible Many bases can be labeled	Difficult to produce Problems with long-term stability

Oligonucleotide	Short sequence — better
	  penetration
	Efficient hybridization
	Chemically synthesized
	Readily produced from
	  sequence databases
	No cloning expertise
	  required
		Limited amount of labeling
		  for each oligomer
		Initial synthesis of cocktails
		  expensive
		Insensitivity of single probes

gene expression in tissue sections. Such probes detect RNA targets which are usually present in many copies per cell unlike genes which are single-copy DNA sequences located on chromosomes.

As a prelude to possible extensive investigations of gene expression, it was thought necessary to prove that RNA was preserved in archived fixed tissues. The lability of RNA to endogenous endonuclease activity is well known, and molecular biologists usually take elaborate precautions to protect RNA from these enzymes. The approach we chose was based on knowledge that virtually all mRNA molecules have a polyadenylated tail. Accordingly, poly-dT probes were synthesized and labeled with biotin in our initial experiments (Pringle et al., 1989). Detection of bound probe was by streptavidin-alkaline phosphatase with final detection of the alkaline phosphatase by fast red. This technique has proved to be extremely useful. We were able to confirm excellent mRNA preservation in our initial experiments (Talbot et al., 1989) in a wide variety of archival human tissues including tonsils (Figure 8.3), breast, and gut. We would now recommend that all studies of RNA by ISH should include an assessment of mRNA preservation in the tissue section by poly-dT probing. In general when tissue morphology appears good, RNA preservation is usually excellent. When the tissue preservation appears poor, then usually RNA signal is drastically reduced. There are, however, exceptions and RNA preservation can be surprisingly good from the most unlikely of sources including post-mortem tissue (Shorrock et al., 1991).

The most important consideration is that fixation appears to inactivate rapidly endogenous endonuclease activity. In the case of autopsy material it seems that in certain instances the nucleases themselves may be more rapidly degraded than the RNA, which they might otherwise attack. In one of our

FIGURE 8.3.   This paraffin-embedded section of formalin-fixed human tonsil was probed using a poly-dT oligonucleotide to detect cytoplasmic mRNA by ISH. The result is striking, with intense staining of many cells within the germinal center of the lymphoid follicles and also of active cells in the surrounding T-cell rich areas. (NBT-BCIP magnification × 160.)

investigations we were even able to show preservation of insulin mRNA in macerated stillbirths (Benton and McKeever, 1995).

This technique of total mRNA detection has many possible applications. It is perhaps most important as a control procedure for all ISH experiments in which mRNA in tissue sections is to be demonstrated. It should constitute a control preparation in all instances where a new probe is to be assessed in a tissue section. In every batch of slides taken in any ISH investigation, the poly-dT probes should be run on every case analyzed. When no RNA is detected then interpretation of ISH results from such material will usually be negative and certainly unreliable.

Having established that mRNA is well preserved in a high proportion of all human archived tissues, further applications were developed. Before describing some of these in more detail it is relevant at this stage to summarize the advantages of using nonradioactively labeled oligonucleotide probes for ISH in the molecular pathology laboratory.

A. SENSITIVITY

Comparative experiments were performed in which we compared the use of synthetic oligonucleotide probes with single-stranded riboprobes and double-stranded DNA probes. In our hands synthetic oligonucleotides produced consistently the best results. The amount of label which can be attached to the 30-mer oligonucleotides, which we use routinely, is clearly substantially less than that which might theoretically be attached to a much larger cloned probe of a kilobase or larger DNA. Therefore, the apparent sensitivity of our system requires explanation. The key elements are the use of cocktails of oligonucleotides often collectively spanning a large portion of the gene, careful selection of probes from available sequence data to provide maximum specificity, improved labeling of probes (see below), and the use of alternative haptens to biotin, such as Digoxigenin and FITC.

Hybridization kinetics are such that probe penetration and subsequent hybridization to target sequences will in most cases be much better with small probes. When a cocktail of perhaps 5 to 15 such probes is applied simultaneously, the end result is a high level of hybridization to target nucleic acid sequences. Further improvements may be related to the following developments.

B. PROBE PREPARATION AND LABELING

Although in theory, probes could be selected from any of the published nucleic acid sequences, in practice considerable care is required with probe selection (Pringle et al., 1990). It is important that the probes should not cross-hybridize to other related or unrelated sequences in the cells under study. It is our practice that all new probes should be screened on RNA extracted from appropriate tissues or cell lines by Northern blotting. This serves the dual purpose of giving an indication of the copy number of the target sequence and also of establishing suitable control tissues for the actual NISH studies. Similar information might sometimes be available from the literature, which may also indicate whether or not probes are likely to work in a variety of different animal species.

Our synthetic oligonucleotide probes are synthesized at a base length of 30 using standard phosphoramidite chemistry. Base additions are made using o-(2-cyanoethyl)-N, N-diisopropylphosphoramidites (from Cruachem, U.K.) (Beaucage and Caruthers, 1981), with subsequent addition of a 5′ primary amine group by reaction with Aminolink 2 (from Applied Biosystems, USA, 400 498) terminal deoxynucleotidyl transferase. Following purification of the oligomers, a TdT reaction (Deng and Wu, 1981) is used to add a homopolymer tail of nucleotide analogs at the 3′ end. Improved homopolymer tailing utilizes TdT (Kendall et al., 1991) and 5-(3-aminoallyl)2′-deoxyuridine 5′ triphosphate (AA-dUTP) (from Sigma, U.K., A 5910). Labeling the aminoallyl and Aminolink oligonucleotides at the 5′ and 3′ ends is achieved with primary amine directed acylating agent Digoxigenin-NHS (from Boehringer). Finally, the labeled oligonucleotides are purified by gel exclusion in Sephadex G50 spun columns (×2) and stored in aliquots of 1 µg at 20°C.

This procedure is considerably more complex than some of the earlier labeling procedures using biotin. Successful results with low copy number nucleic acid targets demand efficient probe labeling. The failure of some research groups to obtain satisfactory results with nonradioactively labeled synthetic oligonucleotides is probably related to the use of single probes or to inefficient labeling procedures.

C. ALTERNATIVE HAPTENS

As explained earlier, many previous studies have used biotin-labeled probes with a subsequent labeled streptavidin or avidin detection system. Such detection systems can be made extremely sensitive, as there is a high affinity between biotin and streptavidin. In attempting to localize hepatitis viruses in the liver using biotin-labeled probes, it quickly became apparent that background staining as a result of endogenous biotin within the liver was a very major problem which rendered most of the results uninterpretable. Biotin levels in tissues are closely related to pyruvate carboxylase activity, since it acts as a carrier of activated CO_2. This inevitably implies that most metabolically active tissues will tend to contain relatively high levels of biotin. Tissues such as the liver, kidney, muscle, and a wide variety of neoplasms are all likely to give high levels of background staining with biotin-streptavidin detection systems. Unfortunately it is these same metabolically active tissues which are most likely to be of interest to tumor pathologists and toxicologists alike.

An ideal hapten for labeling oligonucleotides should have no endogenous equivalent and should also be one with high affinity antibodies directed against the antigen. Digoxigenin and fluorescein are two such haptens (Pringle, 1994), and both can give excellent results. Digoxigenin is probably the most popular hapten in NISH techniques. With alkaline phosphatase detection systems, background activities are usually so low that maximum sensitivity can be achieved by prolonged incubation at the detection phase. The standard method in most laboratories uses 5-bromo-4 chloro-3-indolyl phosphate (BCIP) as the enzyme substrate and nitroblue tetrazolium (NBT) as the chromogen. The amount of stained product is directly proportional to time so that overnight incubations become possible without significantly increasing background.

D. NONRADIOACTIVE LABELING AND DETECTION

Isotope labeling (e.g., ^{32}P or ^{35}S) does provide a very sensitive detection system (Lo, 1986). There are, however, two major difficulties. The first is the relatively poor spatial resolution of many of the isotopic labeling systems often also associated with long-time intervals to allow autoradiographs to develop. Often with ^{32}P labeled probes, the silver grains produced may totally mask the target cell, rendering cytological identification impossible. This may not be a problem with a homogeneous population of tumor cells, but for tissues such as lymph nodes or endocrine organs the value of positive ISH staining is considerably limited if the underlying cell cannot be identified. The second difficulty is that associated with the use of radioactivity. The practical difficulties are all fairly well known, but include limited shelf life of the labeled probe depending on the half-life of the isotope and, of more practical concern, safety considerations including disposal of waste material. For a laboratory proposing a major project involving ISH studies on many slides and tissues, the safety implication may be such that the experiments can only be undertaken using nonradioactively labeled probes.

It is our belief that the latest technological innovations in probe preparation, probe labeling, and detection systems can now give ISH systems which are comparable to radioactive systems in terms of sensitivity, and certainly far superior in terms of spatial resolution. An additional bonus is that the nonradioactively labeled probes also appear to be highly stable with a shelf life of at least 12 months. This has also facilitated the availability of labeled probes from a number of commercial sources, as such probes are readily distributed on a worldwide basis.

IV. IMMUNOGLOBULIN LIGHT CHAIN mRNA AS A MODEL NISH SYSTEM

We selected NISH detection of immunoglobulin mRNA as a model system (Pringle et al., 1990) because of a personal interest in the lymphoid system and its malignancies, and secondly because the range of B cells present in a normal tonsil give an excellent indication of the sensitivity and specificity of the system. Normal tonsil is readily available, and fixed tissue shows excellent morphological preservation of plasma cells (present in large numbers beneath the epithelium) together with lymphoid follicles which are the centers of B cell proliferation and differentiation. This tissue is a very good model system because the light chain mRNA copy number present in the various B cells varies depending on the degree of differentiation (Close et al., 1990).

The undifferentiated mantle zone B lymphocyte contains only low copy number (c. 10 to 20 per cell), whereas immunoglobulin secretory plasma cells contain high copy number (c. 1000+ per cell). Other cells within the germinal centers contain intermediate copy numbers. The ultimate challenge was to develop a system which enabled immunoglobulin mRNA to be identified in the lymphocyte mantle zone, which can now be achieved. We have now used this approach to demonstrate light chain restriction (still the "gold standard" for B-cell malignancy) in several hundred samples of lymphoid tissue (Figure 8.4) (Colloby et al., 1994). One unexpected benefit of the ISH technique (after demonstration of the protein product by immunocytochemistry) was the finding of several cases of myeloma and malignant lymphomas in which there was no detectable light chain protein by immunocytochemistry but in which light chain mRNA could clearly be demonstrated. Further investigations of this interesting phenomenon are still in course, but the discrepancy presumably reflects some form of post-transcriptional defect resulting in a defect in translational activity. The protein is either not produced or produced in a form which is no longer recognizable by the monoclonal antibodies used in the immunocytochemical studies. The possible relevance of these observations to other aspects of tumor biology and toxicology are fairly obvious. The basic methodology developed by us in this model system could be applied, for example, to studying iso-enzyme changes in liver and other tissues following exposure to toxic agents, with the possibility of assessing both transcription and translation by a combination of ISH and immunocytochemistry. Based on our experience with the model system described above, we were able to refine several aspects of the methodology, and these are summarized in Table 8.2.

V. CHOICE OF APPROPRIATE CONTROLS

As in any scientific investigation, the selection of controls is an important part of the methodology. Possible controls for NISH are illustrated in Table 8.3. In addition to these ISH controls, when immunocytochemical detection systems are used, the standard immunocytochemistry controls, such as omission of the labeled antibody, should also be performed.

VI. APPLICATIONS OF NISH TECHNOLOGY

A. CELL PROLIFERATION AND HISTONE mRNA
There are literally hundreds of possible applications of our approach to NISH. One in particular will be described in more detail as it is likely to be of considerable value in many toxicology experiments. As stated above, a particu-

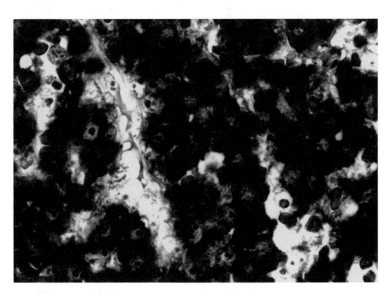

FIGURE 8.4.   This section of fixed high-grade malignant tissue has been probed for immuno-globulin λ-light chain mRNA by ISH. In this instance all of the tumor cells show strong expression of only the λ-light chain message, and probes for K-light chain mRNA showed a positive result only in a few non-neoplastic residual plasma cells. The demonstration of light chain restriction confirms malignancy. Such a result is much more difficult to achieve by conventional immuno-cytochemistry. (NBT-BCIP magnification × 1000.)

TABLE 8.2
Summary of NISH Methodology

Stage	Comments
1. Tissue fixation	Standard aldehyde fixatives usually good
	Carnoy's poor
	Avoid prolonged (>48 hours) fixation
2. Section cutting	Essential to precoat slides with an adhesive. Best results with 3-aminopropyl-triethoxysilane
3. Solution preparation	Use diethyl pyrocarbonate (DEPC) water to ensure RNA preservation
	Prepare RNase-free glassware
4. Pretreatments	Removes cross-linked protein to increase probe penetration to target
(a) Proteinase K normally 2–50 µg/ml for 1 hour at 37°C	Optimum concentration determined empirically as dependent on fixation
(b) Acetylation	Sometimes necessary to block binding of probes to charged groups
5. Hybridization	Important to check that basic methodology is sound for each experiment. Use of controls important.
	Prehybridization steps may be avoided by dehydrating sections prior to hybridization and rehydrating with probe containing hybridization buffer
	Nonspecific binding of probe is blocked by incorporation of macromolecules into hybridization buffer
	Probe concentration normally optimum at 50 ng/oligomer/ml
6. Detection stage	Best results usually obtained with alkaline phosphatase label using BCIP/NBT and overnight incubation if necessary

For more detailed information, see Pringle, J. H., Non-isotopic detection of RNA *in situ.,* in *Non-Isotopic Methods in Molecular Biology: A Practical Approach,* Practical Approach Series, IRL Press, Oxford University Press, Ed., Levy, Chapter 7, 1995, 85.

TABLE 8.3
Controls for NISH

Controls	Comments
1. Negative probes	Use corresponding sense oligomer
	Random sequence oligomer probes
	Use of another antisense probe not expressed in the tissue under study
2. Stringency controls	May be important to eliminate nonspecific hybridization
3. RNase pretreatment	Should abolish the positive signal
	Must ensure that residual enzyme does not inactivate probe
4. Oligo d(T) probes	Confirms RNA preservation in tissue/cell-line
5. Known negative or positive control tissue	Important to check that basic methodology is sound for each experiment
6. Enzyme pretreatment controls	Fixation will affect the optimum length of enzyme digestion
	Use multiple concentrations if necessary
7. No hybridization	Omit probe and perform detection system

lar advantage of NISH technology is the widespread applicability to archived fixed tissue samples. Of particular interest in this context are studies of cell proliferation and growth control. Until recently, counts of mitotic figures were the only way of assessing cell proliferation in tissue sections. This is a time-consuming procedure with many inaccuracies (Quinn and Wright, 1990) due in part to heterogeneity in different parts of tissues and tumors and also to variations in the length of time spent in mitosis as well as in the length of the cell cycle. A more precise measure of cell proliferation is the so-called labeling index in which labels such as tritiated thymidine or 5-bromo-2'-deoxyuridine (Brd Urd) are used to label cells in the DNA S phase. Such an approach requires pre-exposure of the cells under study to the appropriate labeled nucleotide. In human tissues this has only rarely been undertaken and in any event cannot be undertaken retrospectively in archived tissues.

Alternative approaches include immunocytochemical demonstration of a variety of antigens which are differentially expressed in proliferative cells. The two most widely used antibodies have been Ki67 (Gerdes et al., 1992) and proliferating cell nuclear antigen (PCNA) (Waseem and Lane, 1990). Ki67 unfortunately is not well expressed in fixed tissues, and PCNA appears to have a long half-life, so that a high proliferation index obtained with this antigen may indicate that cell proliferation has taken place without necessarily providing information on a labeling index at discrete time points.

The histone proteins H2b, H3, and H4 are produced during the S phase of the cell cycle (Chou et al., 1990) and preceded by the expression of high copy mRNA (Zhong et al., 1983). To facilitate studies of histone mRNA, a cocktail of Digoxigenin-labeled oligonucleotides was prepared with three 30 mers each for H2b, H3, and H4, respectively. These were tested initially in normal tonsil where they showed clearly the proliferating cells in the general centers of the lymphoid follicles and also the basal cells of the tonsil crypt epithelium. In collaboration with the Royal Postgraduate Medical School, London, experiments were undertaken to compare, in the rat partial hepatectomy model, the labeling index as assessed by histone mRNA detection with that measured by Brd Urd in the same animals (Figure 8.5) (Alison et al., 1994). A graph was plotted to show the relationship of the two measurements and a very high correlation coefficient suggested near homology of the two quite separate approaches. We have subsequently confirmed the general applicability of this methodology in a wide variety of human and animal tissues, including archived samples of 30 or more years. In animals subjected to previous toxicity studies, a precise labeling index can now be obtained retrospectively.

B. DETECTION OF APOPTOSIS

Our studies of lymphoma tissues using histone mRNA expression showed a high level of proliferation in high-grade compared to low-grade malignant lymphomas (Figure 8.6). Another important factor in growth control in lymphoma tissues is the rate of cell death as determined by apoptotic activity. This is extremely difficult to detect by conventional light microscopy, as the

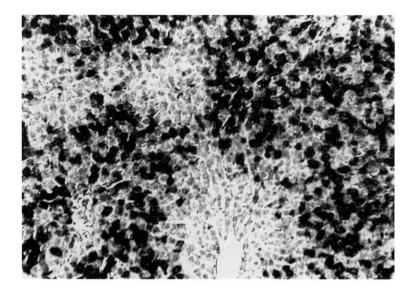

FIGURE 8.5. The illustration shows the histological appearance of a rat liver 24 hours after a two-thirds partial hepatectomy. The section has been probed for histone 2b, 3, and 4 mRNA, and the labeled cells can be clearly identified in the periportal areas. At this stage, most of the labeled cells are hepatocytes. This corresponds to intense proliferative activity in these cells as a result of active regeneration. (NBT-BCIP magnification × 400. Illustration courtesy of Dr. Malcolm Alison.)

classic apoptotic bodies are often only visible at the ultrastructural level. Apoptosis is accompanied by progressive fragmentation of DNA, giving rise to the classic pattern of "ladders" visible after electrophoresis. The formation of breaks in DNA at an early stage of apoptosis offers an opportunity to demonstrate apoptotic cells using a modification of the nick translation method used to label cloned DNA probes. If Digoxigenin-labeled (Digoxigenin-11-dUTP) nucleotides are applied to sections together with the enzyme, terminal deoxynucleotidyl transferase (TDT), then this *in situ* end labeling (ISEL) method, combined with an alkaline phosphatase detection system, will demonstrate apoptotic cells as illustrated in Figure 8.7. We have also been able to show that the method can be adapted for use in suspensions, thus allowing quantification by analytical flow cytometry.

C. OTHER TECHNIQUES

As shown above, NISH has much to offer in studies of gene expression in a variety of archived tissue samples in tumor biology and toxicology studies. Another major interest is the possibility of detecting cytogenetic changes in tissue sections. Conventional cytogenetics on pathology samples is often either impossible (specimen already fixed) or extremely difficult because of the necessity to culture cells and generate metaphase spreads. In many instances it may not be certain that any cytogenetic changes observed are representative of the tumor or tissue studied, since both can exhibit considerable heterogeneity.

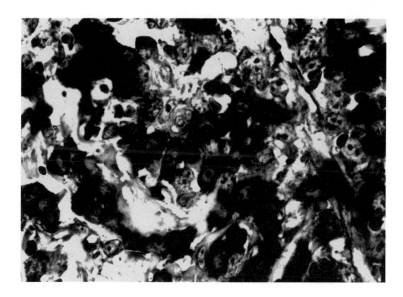

FIGURE 8.6.   This is the same high-grade malignant lymphoma illustrated in Figure 4. In this section a cocktail of oligonucleotides was used to detect histone mRNA for H2b, H3, and H4. The numerous darkly staining cells are easily identified. In separate experiments two- to threefold lower expression was found in low grade lymphomas, indicating a lower level of cell proliferation in the latter group. (NBT-BCIP magnification × 1000.)

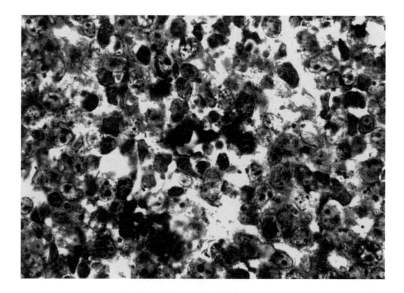

FIGURE 8.7.   This is an illustration of a lymphoid tumor in which apoptotic cells have been demonstrated by the *in situ* end labeling technique (ISEL). Note the presence of the occasional darkly stained apoptotic cells. Such cells are often extremely difficult to identify in conventional histological sections. (NBT-BCIP magnification × 1000.)

The use of labeled probes (for example fluorescent probes, thus FISH: *F*luorescent *I*n *S*itu *H*ybridization) targeted against highly repetitive DNA sequences in theory facilitates a wide variety of so-called interphase cytogenetic structures. Probes are available which will identify most human chromosomes and gross changes such as trisomies. More subtle chromosomal changes are often detectable by PRINS (*Primed I*n *S*itu) in which chromosomes are labeled *in situ* (Koch et al., 1989). After denaturation of chromosomal DNA, designed oligonucleotides are first allowed to hybridize and are then extended with labeled deoxynucleotide triphosphate, using DNA polymerase. Signal intensity limits the method somewhat, and in our hands it is most effective with repetitive DNA sequences or with high copy mRNA following reverse transcriptase. These methods, however, are still at the developmental stage, and further enhancements in sensitivity can be anticipated.

Double labeling techniques which combine immunocytochemistry with NISH represent a further enhancement of the technology. This allows simultaneous immuno-identification of a cell type together with an assessment of the expression of a particular gene in the localized cells. We have found this easiest to achieve using fluorescent probes and antibodies. The technique was originally used by us to examine J-chain gene expression in IgA containing plasma cells (Harper et al., 1991), but we have also used it to study Epstein-Barr virus and p53 in AIDS-related malignant lymphomas.

VII. CONCLUSION

The techniques described in this article will permit toxicologists to examine, retrospectively, toxic effects of drugs and chemicals on a wide variety of animal tissues by studying alterations in genes and gene expression. The archives of tissue maintained in most pathology laboratories are a precious resource which must be protected from pressures to discard the material as and when laboratory space becomes more and more constrained. Further developments in technology will no doubt increase the value of the archived tissues and provide yet more exciting research opportunities for the next generation of molecular pathologists.

REFERENCES

Alison, M., Chaudry, Z., Lauder, I., and Pringle, H., Liver regeneration: *in situ* hybridization for histone in RNA with bromodeoxyuridine labeling for the detection of S-phase cells, *J. Histochem. Cytochem.*, 42, 12, 1603, 1994.

Beaucage, S. L. and Caruthers, M. H., Deoxynucleoside phosphoramidites — a new class of key intermediates for deoxypolynucleotide synthesis, *Tetrahedron Lett.*, 22, 1859, 1981.

Benton, K. and McKeever, P., Pancreatic hormonal assessment in relation to fetal intrauterine growth retardation, (submitted for publication), 1995.

Brocksmith, D., Angel, C. A., Pringle, J. H., and Lauder, I., Epstein-Barr viral DNA in Hodgkin's Disease: amplification and detection using the polymerase chain reaction, *J. Pathol.*, 165, 11, 1991.

Chou, M. Y., Chang, A. L. C., McBride, J., Donoff, B., Gallagher, G. T., and Wong, D. T. W., A rapid method to determine proliferation patterns of normal and malignant tissues by H3 mRNA *in situ* hybridization, *Am. J. Pathol.*, 136, 729, 1990.

Close, P. M., Pringle, J. H., Ruprai, A. K., West, K. P., and Lauder, I., Zonal distribution of immunoglobulin — synthesising cells within the germinal centres: an *in situ* hybridisation and immunohistochemical study, *J. Pathol.*, 162, 209, 1990.

Colloby, P. S., West, K. P., Pringle, J. H., Hansmann, M., Hell, K., Lorenzen, J., Hemming, D., Lauder, I., Light chain restriction in B-cell lymphoproliferative disorders — *in situ* hybridisation for mRNA is a viable alternative to immunocytochemistry, *G. Pathol.*, (in press), 1995.

Deng, G. and Wu, R., An improved procedure for utilizing terminal transferase to add homopoly-mers to the 3′ termini of DNA, *Nucleic Acids Res.*, 9, 4173, 1981.

Gall, J. G. and Pardue, M. L., Formation and detection of RNA-DNA hybrid molecules in cytological preparations, *Proc. Natl. Acad. Sci. U.S.A.*, 63, 378, 1969.

Gerdes, J., Becker, M. H. G., and Key, G., Immunohistological detection of tumour growth fraction (Ki-67 antigen) in formalin fixed and routinely processed tissues, *J. Pathol.*, 168, 85, 1992.

Harper, S. J., Pringle, J. H., Gillies, A., Allen, A. C., Layward, L., Feehally, J., and Lauder, I., Simultaneous *in situ* hybridization of mRNA and immunoglobulin detection by conven-tional immunofluorescence in paraffin embedded sections, *J. Clin. Pathol.*, 45, 114, 1991.

John, H. A., Birnstiel, M. L., and Jones, K. W., RNA-DNA hybrids at the cytological level, *Nature*, 223, 582, 1969.

Kawasaki, E. S., The polymerase chain reaction: its use in the molecular characterization and diagnosis of cancers, *Cancer Invest.*, 10(5), 417, 1992.

Kendall, C. H., Roberts, P. A., Pringle, J. H., and Lauder, I., Parathormone mRNA detection by *in situ* hybridization in normal and abnormal parathyroid tissue, *J. Pathol.*, 165, 111, 1991.

Koch, J., Kolvraa, S., Petersen, K. B., Gregersen, N., and Bolund, L., Oligonucleotide-priming methods for the chromosome-specifiic labelling of alpha satellite DNA *in situ, Chromosoma*, 98, 259, 1989.

Koopmans, M., Monroe, S. S., Coffield, L. M., and Zaki, S. R., Optimization of extraction and PCR amplification of RNA abstracts from paraffin-embedded tissue in different fixatives, *J. Virol. Meth.*, 43, 189, 1993.

Libetta Carole, Pringle, J. H., Angel Carole, Craft, A. W., Malcolm, A. J., and Lauder, I., Demonstration of Epstein-Barr viral DNA in formalin fixed, paraffin embedded samples of Hodgkin's Disease, *J. Pathol.*, 161, 255, 1990.

Lo, C. W., Localisation of low abundance DNA sequences in tissue sections by *in situ* hybridiza-tion, *J. Cell. Sci.*, 81, 43, 1986.

Pringle, J. H., Isotopic or chromogenic? A comparative appraisal of probe labelling for *in situ* hybridisation, *Proc. Royal Soc. Microscopol. Sci.*, (Micro 90), 695, 1990.

Pringle, J. H., Non-isotopic detection of RNA *in situ,* in *Non Isotopic Methods in Molecular Biology: A Practical Approach*, Ed., Levy, Practical Approach Series, IRL Press, Oxford University Press, Chapter 7, 1995, 85.

Pringle, J. H., Lindsay Primrose, Kind, C. N., Talbot, I. C., and Lauder, I., *In situ* hybridisation demonstration of polyadenylated RNA sequences in formalin-fixed paraffin sections using a biotinylated oligonucleotide polyd(T) probe, *J. Pathol.*, 158, 279, 1989.

Pringle, J. H., Ruprai, A. K., Potter, L., Kyte, J., Close, P. M., and Lauder, I., *In situ* hybridization detects insulin and glucagon secreting cells in routinely fixed and processed pancreatic tissue, *J. Pathol.*, 162, 197, 1990.

Quinn, C. M. and Wright, N. A., The clinical assessement of proliferation and growth in human tumours: evaluation of methods and applications as prognostic variables, *J. Pathol.*, 160, 93, 1990.

Shorrock, K., Pringle, J. H., Roberts, P., and Lauder, I., *In situ* hybridization detects insulin and glucagon secreting cells in routinely fixed and processed pancreatic tissue, *J. Pathol.*, 165, 105, 1991.

Talbot, I. C., Primrose Lindsay, Pringle, J. H., Functional activity of intestinal epithelium demonstrated by mRNA *in situ* hybridisation, *J. Pathol.*, 158, 287, 1989.

Turner, P. C., Bailey, A. S., Cooper, R. J., and Morris, D. J., The polymerase chain reaction for detecting adenovirus DNA in formalin-fixed, paraffin-embedded tissue obtained *post mortem*, *J. Infect.*, 27, 43, 1993.

Volkenandt, M., Koch, O., Fanin, R., Banerjee, D., Seger, A., Vogel, J., Bierhoff, E., Heidl, G., Neyses, L., and Bertino, J. R., Sequence analysis of polymerase chain reaction amplified t(14;18) chromosomal breakpoints in formalin fixed, paraffin wax embedded follicular lymphoma, *J. Clin. Pathol.*, 45, 210, 1992.

Warford, A. and Lauder, I., In situ hybridisation in perspective, *J. Clin. Pathol.*, 44, 177, 1991.

Warford, A., Pringle, J. H., Hay, J., Henderson, D., and Lauder, I., Southern blot analysis of DNA extracted from formol-saline fixed and paraffin wax embedded tissue, *J. Pathol.*, 154, 313, 1988.

Waseem, N. H. and Lane, D. P., Monoclonal antibody analysis of the proliferating nuclear antigen (PCNA), *J. Cell Sci.*, 96, 121, 1990.

Young, B. D., Molecular biology in medicine, *Postgrad. Med. J.*, 68, 251, 1992.

Zhong, R., Roeder, R. G., and Heintz, N., The primary structure and expression of four cloned human histone genes, *Nucleic Acids Res.*, 11, 7409, 1983.

Chapter 9

GENETIC VARIATION IN HUMAN AND LABORATORY ANIMAL POPULATIONS AND ITS IMPLICATIONS FOR TOXICOLOGICAL RESEARCH AND HUMAN RISK ASSESSMENT

Michael F.W. Festing

CONTENTS

I. Response to Many Xenobiotics is under Genetic Control 166

II. Genetic Structures of Human and Laboratory Animal
Populations Differ 168
   A. Genetic Structure of Human Populations 168
   B. Genetic Structure of Laboratory Animal Populations 168
      1. Outbred Stocks 169
      2. Inbred Strains 170
      3. Different Genetic History of Laboratory Mice and Rats 170

III. Toxicological Screening and Research Uses Animals of a
Narrow Range of Genotypes 171

IV. Strain Differences as a Research Tool 173
   A. The Type of Genetic Variation Differs between Humans
and Laboratory Animals 173
   B. Failure to Develop the Science of Pharmacogenetics 174
   C. Failure to Develop Adequate Animal Models 174

V. The New Genetics and Its Relevance to Toxicological Research ... 175

VI. Better Experimental Design Could Accommodate More than
One Strain at Little Extra Cost 176
   A. High Precision in an Experiment 176
   B. Increasing the Range of Applicability of an Experiment 178

VII. Conclusion 180

References 181

0-8493-9229-2/95/$0.00+$.50
© 1995 by CRC Press Inc.

I. RESPONSE TO MANY XENOBIOTICS IS UNDER GENETIC CONTROL

There is good evidence that the response to many toxic chemicals has a strong genetic component. In humans about 50 Mendelian loci have been identified which are associated with chemical hypersensitivity (Calabrese, 1991). Some of these are common in all human populations. Some polymorphisms associated with resistance to malaria, including sickle cell anemia/trait, the thalassemias, and glucose-6-phosphate dehydrogenase (G6pd) deficiency are particularly common in tropical countries and are associated with increased sensitivity to several chemicals. It is estimated that 400 million people suffer from G6pd deficiency, with 300 different variants of the protein known (Vulliamy et al., 1992). Other common polymorphisms include slow vs. fast acetylation, first detected by the adverse reaction to isoniazid, and that concerning the metabolism of debrisoquine, which is associated with adverse response to several drugs as well as "spontaneous" tumors (Idle et al., 1992). These polymorphisms have been reviewed recently by Calabrese (1991) and Bochkov and Titenko (1991).

The response to many xenobiotics has a polygenic mode of inheritance, i.e., one where sensitivity is the product of polymorphism at several loci capable of interacting with environmental factors (Kalow, 1991). In humans, it is difficult to separate familial effects which are due to genetics and those due to environmental effects common to the whole family, so polygenic inheritance is most clearly documented as a high level of concordance among monozygous twins (especially if reared apart), or as a familial sensitivity which can be separated from identifiable environmental effects. Susceptibility to lung, breast, and other forms of cancer probably has this mode of inheritance, though some individual polymorphic loci are now beginning to be identified. For example, variation in glutathione-S-transferase is associated with increased susceptibility to tobacco smoke-induced lung cancer (Kihara et al., 1994). Experimentally, Vesell and Page (1968a–c) have shown that the rate of elimination of drugs such as dicoumarol, phenylbutazone, and antipyrine is highly inherited with a polygenic mode of inheritance. However, because of the difficulties of studying twins, and because it would not be ethical to subject them to large doses of toxic chemicals, such variation is difficult to identify and may be severely underestimated.

Genetic factors are of obvious significance with respect to exposure to industrial chemicals and to medicinal compounds, where adverse reactions to drugs may be seen in people with genetically determined sensitivity. The World Health Organization recommends that "the limits for environmental and occupational exposure to hazardous chemicals should protect all individuals, including those who may be hypersusceptible. Procedures to achieve this goal need to be developed" (Grandjean, 1991).

In laboratory animals, there is also substantial genetic variation in response to xenobiotics though fewer Mendelian loci have been identified in mice and rats than in humans (Festing, 1987) even though several million mice and rats are used each year in toxicity testing. This is partly a reflection of the different population structures of humans and animals (see below), and partly because genetic variation in response is only detectable if the relationships among affected individuals are known. The relationships among mice or rats on a toxicity trial are never known, as it is impractical to record the pedigree of each individual. Usually the only way of discovering that the response is under genetic control is to study more than one strain of the species, which is rarely done.

Strain differences in response to xenobiotics are almost universal. They include differences in acute toxicity, neurotoxicity, carcinogenesis, teratogenesis, and immunotoxicological reactions. Some examples taken almost at random are given in Table 9.1.

TABLE 9.1
Examples of Strain Differences in Response to Xenobiotics

Xenobiotic	Species	Comment	Reference
Butylated hydroxytoluene	Mouse	Large strain differences in LD_{50} ranging from 138 mg/kg in strain DBA/2 to 1739 mg/kg in BALB/c	Kawano et al. (1981)
Cocaine	Mouse	Significant strain, age, and sex effects on hepatotoxicity in a test involving 5 strains	Smolen and Smolen (1990)
ENU	Mouse	Growth of preneoplastic hepatic lesions is quicker in susceptible C3H mice than in resistant C57BL/6 mice	Hanigan et al. (1988)
Urethane	Mouse	Strain differences in susceptibility to lung tumors, with strain A/J susceptible and C57BL/6 resistant	Malkinson et al. (1985)
3,2'-dimethyl-4-aminobiphenyl	Rat	Carcinomas of prostate found in 50% of F344 rats, in lower percentages in three other strains and in none of the Wistar strain	Shirai et al. (1990)
Acetaminophen	Rat	Nephrotoxicity: SD strain rats are resistant, F344 susceptible possibly due to strain differences in metabolism	Newton et al. (1985)
Diisopropyl fluorophosphate	Rat	Neurotoxicity: Strain F344 more resistant than SD or LE outbred rats	Gordon and MacPhail (1993)
DMBA	Rat	Mammary carcinogenesis: large differences in numbers of tumors per rat among 10 rat strains with strain COP being entirely resistant	Isaccs (1988)

II. GENETIC STRUCTURES OF HUMAN AND LABORATORY ANIMAL POPULATIONS DIFFER

Although laboratory animals are widely used as "models" of humans, the genetic structures of human and laboratory animal populations are quite different.

A. GENETIC STRUCTURE OF HUMAN POPULATIONS

Most human populations are extremely large, with a pool consisting of thousands or even millions of individuals potentially able to marry and contribute to the following generation, though some cultural and/or geographical barriers to the free flow of genes remain. For example, Rodriguez-Larralde et al. (1994) found a high negative correlation between coefficient of kinship and geographical distance among people in Sicily. However, there is usually substantial flow of genes between populations, particularly in recent history. Thus in countries such as the United States, migrants from many different races and isolated groups are mixing in what will eventually be a single population with few geographical or cultural barriers to the flow of genes between individuals. Integration of migrant populations is proceeding faster than might have been expected. For example, Coleman (1994) notes that up to 40% of West Indians born in the United Kingdom appear to have white partners. Close inbreeding is usually avoided, though throughout North and sub-Saharan Africa and West, Central, and South Asia a large proportion of marriages are preferentially between close biological kin, particularly among those of low socioeconomic status (Shami et al., 1994). Other complexities include possible racial differences in speed of migration and assortative mating in which mates are chosen on the basis of a high similarity phenotypically and culturally (Biondi et al., 1993).

The large population size, assortative mating, and low levels of inbreeding tend to lead to strong conservation of genetic variation (Falconer, 1981, Chap. 2). Mutant alleles, even if quite unfavorable, tend to be maintained in the population for many generations, particularly if they have a recessive mode of inheritance. In this case, selection against the homozygous individual is quite ineffective in lowering the frequency of the gene, since most alleles are maintained in a heterozygous state. Thus, unfavorable genes, such as those causing cystic fibrosis and muscular dystrophy, remain in the population for many hundreds of generations, if not indefinitely.

B. GENETIC STRUCTURE OF LABORATORY ANIMAL POPULATIONS

The genetic structure of laboratory animal populations is entirely different from that of humans. In general, populations of laboratory mice and rats are divided into small noninterbreeding groups with rigid geographical separation. There are two main sorts of populations: so called "outbred stocks," in which there is no deliberate attempt to inbreed, and "inbred strains," which have been brother × sister mated for many generations.

1. Outbred Stocks

Outbred stocks of laboratory animals are usually closed colonies (i.e., there is no immigration into the colony) which have been maintained without deliberate inbreeding. Typically, they are designated by their historical origin, such as "Wistar" strain rats, which were originally maintained at the Wistar Institute, from which they were distributed to research workers throughout the world. Similarly, "Swiss" mice were developed from nine mice imported into the United States from Lausanne in 1926 by Dr. Clara Lynch, and subsequently widely distributed by L.T. Webster following his work on susceptibility to infectious agents (Rice and O'Brien, 1980). Many of the outbred stocks have been maintained as small breeding colonies for many generations. Sometimes genetic "bottlenecks" have occurred, particularly when population numbers have been reduced to a few pairs when stocks are sent to a new location, or were "cleaned up" through a process of hysterectomy re-derivation to eliminate microbial pathogens. Some colonies of outbred mice and rats have been maintained as very small back-up populations in thin-film isolators as an insurance against a disease outbreak in the larger production colony. This can lead to rapid inbreeding with associated genetic drift, which may be fixed when the main colony is replaced by the backup colony. In some cases, outbred stocks have been deliberately brother × sister mated for a few generations, in order to fix their characteristics. For example, the CFW and CF1 stocks of mice, which have been widely used by toxicologists on the assumption, presumably, that the stocks are genetically variable, have a history of over 20 generations of brother × sister mating in their ancestry (Festing, 1979b).

The levels of inbreeding are occasionally reduced by inadvertent outcrosses (Festing, 1974) which often go undetected, as very few outbred stocks are subjected to genetic quality control procedures.

The net effect is that the actual level of genetic variability in outbred stocks can range from being quite high and virtually equivalent to that found in wild populations (Rice and O'Brien, 1980) to very low levels approaching those found in inbred strains (Festing, 1993a). The actual level for any individual colony is usually not known, unless it is specifically tested using a number of genetic markers known to be polymorphic within the species. Although a wide range of these markers is now available, few outbred stocks used in toxicology research have been so studied.

One important effect of the geographical separation, and associated inbreeding and genetic drift, is that colonies with the same name drift apart genetically, so that, eventually, generic names such as Wistar, Sprague-Dawley, and Swiss become meaningless in terms of biological properties. It is not widely recognized that there are no genetic markers that will distinguish between Wistar and Sprague-Dawley rat stocks, largely because each colony is different (Lovell and Festing, 1982; van Zwieten et al., 1984; Yamada et al., 1979). Different colonies of Sprague-Dawley (SD) stocks can have quite different patterns of life-span and spontaneous disease. For example, Anver et al. (1982) found that one colony of SD rats had a 50% survival of 24 months and maximum survival of 29 months, whereas rats from another colony maintained under identical conditions had only 10% mortality at 24 months and maximum survival of 38 months. Wistar rats can be expected to vary in a similar way.

2. Inbred Strains

Inbred strains are produced by 20 or more generations of brother × sister mating, with all individuals tracing back to a single breeding pair in the 20th or a subsequent generation (to eliminate parallel sublines). The resulting colony can be compared with a clone of genetically identical individuals with a fixed genotype. The effect of this inbreeding is to reduce variability within each strain, increasing, at the same time, variability between strains. Thus, if an outbred stock were to be brother × sister mated for 20 generations with all offspring being kept, the result would be a whole series of separate substrains like the branches of a tree, with little variation within each branch, but substantial variation, well outside the limits found in the original stock, between the different branches (Falconer, 1981, Chap. 15). An inbred strain would be one of these branches, and two inbred strains from the same base stock may differ widely, depending on the amount of genetic variability within the original outbred stock.

There are over 400 inbred strains of mice (Festing, 1993b) and 200 inbred strains of rats (Greenhouse et al., 1990) available throughout the world. These represent a wealth of experimental material which may be used to study genetic variation in response to xenobiotics in these species. Recent developments in genetics (see below) means that it is now possible to identify some of the genes responsible for differences in xenobiotic response among different inbred strains.

3. Different Genetic History of Laboratory Mice and Rats

The origins of laboratory mice and rats are very different, and this may have some bearing on their genetic characteristics. Mice have been domesticated as pet and "fancy" animals since before Roman times both in Europe and the Far East (Festing and Lovell, 1981). Laboratory mice appear to have been derived from these pet mice. Studies of genetic variation in mitochondrial DNA suggest that almost all mice of the common inbred strains are derived from a single breeding female ancestor about 200 years ago (Ferris et al., 1982), and recent studies of the Y chromosome suggest that these strains were derived from two subspecies *Mus musculus musculus* and *M.m. domesticus* (Nishioka, 1987). Studies of genetic variation in the major histocompatibility complex show that these domesticated laboratory mice differ widely from wild mice, and many congenic strains with wild-derived haplotypes have been developed (Klein, 1989). Recently a number of inbred strains have also been established from wild mice of *Mus musculus domesticus* and of various sub-species (Bonhomme and Guenet, 1989), though little is known about their responses to xenobiotics.

In contrast, *Rattus norvegicus* was not known in Western Europe until the middle of the 18th century. The species migrated from the Far East, reaching

Europe early in the 18th century and England between 1728 and 1730. It gradually replaced the native rat *Rattus rattus*. At one time large numbers of rats were used for rat baiting. Wild rats were placed in an arena, and a terrier was let loose among them. Bets were placed on the number of rats that would be killed in a given time. It is thought that the first domesticated rats were derived from spontaneous albino mutants which were trapped when collecting rats for rat baiting (Lindsey, 1979). Apparently, nobody has attempted to domesticate wild *Rattus norvegicus* from wild rats trapped in the Far East, near the center of the origin of the species, where genetic variability would be expected to be greatest. This history suggests that laboratory rats are likely to be slightly more uniform genetically than laboratory mice, though undoubtedly there is still sufficient genetic variation for large differences in response to xenobiotics among animals of different strains and colonies to be observed.

III. TOXICOLOGICAL SCREENING AND RESEARCH USES ANIMALS OF A NARROW RANGE OF GENOTYPES

Most academic toxicological research is conducted on animals with a very narrow range of genotypes. For example, a survey (Festing, 1990) of 83 papers published in *Toxicology and Applied Pharmacology* found that 93% of them had used a single strain of animals. This means that most investigators would have no idea of the extent to which the observed response was strain-dependent. While it is perfectly valid to use a single strain in many experiments, caution is necessary in extrapolating the results to the species as a whole. If the experiment was done only with F344 rats, it would be unwise to conclude, for example, that the given effects would be observed at the same dose level in "rats" rather than "F344 rats." Yet more than 80% of authors were judged to have generalized their results to the species as a whole from work using only a single strain, as judged by their failure to indicate the specific strain either in the title or in the abstract of their paper. As an example of the possible importance of strain differences in response, Pohjanvirta et al. (1988) in the same sample of papers showed that the LD_{50} of TCDD was substantially above 3000 µg/kg in the H/W strain and was about 10 mg/kg in the Long-Evans strain, a difference of more than two orders of magnitude among two rat strains. Failure to appreciate the importance of genetic factors may well be responsible for causing problems of interpretation in toxicological research. For example, lead is known to be highly toxic when absorbed at high levels, and to produce more subtle effects, such as poor learning ability in children, at lower levels (Chisholm, 1984). Mailman and Lewis (1987) noted that "the effects of small doses of lead on CNS development have been difficult to quantify or study mechanistically," and Hammond et al. (1984) noted that "... the response of individual animals [to lead] is extremely variable, both as to the pattern of lead build up and as to the degree of exposure." Yet a computer search of recent literature failed to detect a single paper which had used more than one strain, and virtually all authors used outbred animals of unknown genetic constitution. If the response of some of the more subtle effects of lead is under genetic control, it is hardly surprising that rats from different colonies give different responses. Discussion of the reasons for the variability in response failed even to mention genetic variation among the animals as one possibility (Festing, 1991).

Strain differences represent a research tool which can illuminate toxicological mechanisms (see Section IV). Failure to explore whether such differences exist may mean that other research workers may not get the same results using another strain, or, alternatively, that they may go to considerable expense and inconvenience to obtain animals of the same strain, when in fact the results may not be strain-dependent. A similar situation is found in the screening of chemicals for carcinogenesis. This is *always* done using a single stock or strain. McAuslane et al. (1991) found that about 20 to 24% of repeat-dose and carcinogenesis screening studies used isogenic F344 rats or B6C3F1 hybrid mice, and the rest were done using a single outbred stock of mice or rats. Typically, a carcinogenesis screen such as that done by the National Toxicology Plan Carcinogenesis Bioassay Program (NTP-CBP) uses four dose levels and both sexes. Groups of about 50 animals are given the maximum tolerated dose (MTD) (i.e., the dose which is estimated not to shorten life-span from noncarcinogenic toxicity), a half and a quarter of the MTD, and a zero dose level control. There is no reason why all 50 animals have to be of the same genotype, and indeed, those who use outbred stocks recognize this. However, if the aim is to screen on animals with a relatively wide range of genotypes (as seems to be desirable), it would be statistically and biologically more efficient to subdivide the 50 animals into, say, four groups of 12 animals each of a different strain (Festing, 1987, 1993a). If the test substance is not carcinogenic, then this will not increase the number of false positive results (i.e., it would not increase the chance that a noncarcinogen is labeled as a carcinogen). Likewise, if there are no strain differences in susceptibility, the use of a multi-strain assay of this sort would be exactly equivalent to a single-strain assay. However, when there are strain differences, then a multi-strain assay of this sort will, on average, be more powerful than the single-strain assay. As an example, consider the data of Shellabarger et al. (1978), who found that diethylstilbestrol (DES) caused no extra tumors in SD rats, but about 70% tumors in ACI strain rats, against a background of about 1% tumors. Assume that about half of all strains of rats are like SD rats and the other half like ACI, then, if a single strain is used, there would be a 50% chance of picking a resistant strain and declaring DES to be noncarcinogenic. However, if four strains were used, the probability that at least one sensitive strain was used would increase from 50% to 93.75% $(1-(1/2)^4)$. If one or more sensitive strains were used, on average about 70% of the 12 animals (i.e., about 8 animals) would develop tumors. This would be a statistically significant increase over the background. This is essentially the same argument as that used by people who advocate the use of outbred stocks, but with the advantage that the problems of using such stocks are avoided. The advantages of multi-strain screening protocols of this sort have been argued extensively over a period of several years (Festing, 1975, 1979a, 1980a, b, 1986, 1990, 1993a), and the statistical implications have been examined in detail by Haseman and Hoel (1979) and Felton and Gaylor (1989), who find that such designs would be statistically more powerful than existing single-strain designs.

IV. STRAIN DIFFERENCES AS A RESEARCH TOOL

Given that there is extensive genetic variation in the response of laboratory animals to xenobiotics, such variation should provide a useful additional tool for studying mechanisms of toxicity. Historically, this has proved to be the case, even though this methodological approach has not been as widely used as it might have been. As early as 1962, Kalow published a book on pharmacogenetics which appeared to offer an exciting new approach to the study of toxic mechanisms through studying genetic variations in response. This was followed by extensive work by Nebert and colleagues in the 1970s on the murine *Ah* locus, the cytochrome P450s, and other polymorphic enzymes associated with detoxification and metabolic activation of chemicals, which undoubtedly led to a better understanding of the way in which chemicals are both detoxified and can cause cancer in animals and humans (Nebert, 1980a,b). However, as pointed out by Kleeberger and Levitt (1991) "... the development of well characterized genetic models has not progressed very rapidly." It is not entirely clear why this should be so, though it is probably the result of a combination of several factors, as discussed below.

A. THE TYPE OF GENETIC VARIATION DIFFERS BETWEEN HUMANS AND LABORATORY ANIMALS

In humans, the type of genetic variation that can be studied most easily is the single-gene Mendelian loci associated with adverse reactions to drugs and environmental chemicals. In laboratory animals, few such loci have been found, almost certainly because of the difference in the population structure between humans and laboratory animals, which has been noted above. In order to find Mendelian genes in animals homologous to those found in humans (e.g., glucose-6-phosphate dehydrogenase deficiency, or the drug acetylation polymorphism), it is often necessary to screen many strains and possibly even wild populations, or induce mutations deliberately. Neither approach would be very familiar to most toxicologists. However, both these polymorphisms exist in mice — the first as a result of an irradiation experiment, and the second as a polymorphism found among inbred strains (Green, 1989).

B. FAILURE TO DEVELOP THE SCIENCE
OF PHARMACOGENETICS

For some reason, pharmacogenetics never attracted a critical mass of interested research workers. No journal specializing in it was established until the 1990s, and although there have been several *ad hoc* meetings on the subject, there is no scientific society concerned primarily with the subject, and few academic departments or research institutes specialize in it. The Jackson Laboratory, a cancer research laboratory founded on the use of mouse genetics as a research tool, might have been expected to play an important role in the development of pharmacogenetics. However, their discovery of the viral etiology of mammary tumors in 1933 (Festing, 1979b) stimulated interest in the viral and immunological basis of cancer, with much less emphasis on its chemical induction. In contrast, subjects such as immunogenetics and behavior genetics have had active scientific societies and relevent journals for many years. The slow development of pharmacogenetics is surprising in view of the potential importance of genetic hypersensitivity in humans and animals for the pharmaceutical industry, which might have been expected to offer considerable financial support for a fledgling scientific discipline.

C. FAILURE TO DEVELOP ADEQUATE ANIMAL MODELS

Immunogeneticists and behavior geneticists were also able to take a longer-term view and develop strains of mice and rats specifically for their own research purposes. In the case of immunogenetics, over 800 congenic strains have been developed to study the properties of the mouse major histocompatibility complex and other loci of immunological importance (Klein, 1989). Each strain requires 10 to 12 backcross generations, which will usually take about three years, so it represents a considerable investment in research effort. Such strains are used widely in research (Weir, 1986). Behavior geneticists have often used selective breeding to develop strains differing widely in various types of behavior, including maze-running, "emotionality," and alcohol preference. The LS and SS (long-sleep and short-sleep) mouse strains are of particular interest. These differ widely in sleeping time following intraperitoneal injection of ethanol. Surprisingly, the strains appear to differ not in rate of ethanol metabolism, but in neural sensitivity to ethanol (Phillips et al., 1989). In contrast, only a few strains have been developed specifically for research in pharmacogenetics. There are some congenic strains differing in Ah inducibility (Nebert, 1980a,b) and N-acetylation (Levy et al., 1992), and selective breeding was used to develop the SENCAR mouse strain (Slaga, 1983), which is sensitive to topically applied carcinogens and has been used extensively to study initiation and promotion as separate steps in the carcinogenesis process. However, this is just a handful of strains, compared with the hundreds used in immunogenetics.

The development of sets of recombinant inbred strains by crossing standard inbred strains and inbreeding to develop a number of parallel strains (Bailey,

1971; Taylor, 1989) provided new research tools which made it possible in some cases to identify individual loci controlling strain differences in response to xenobiotics. Although they are used by a few toxicologists (e.g., Malkinson et al., 1985), few laboratories outside the United States had the resources to maintain and use these strains effectively.

V. THE NEW GENETICS AND ITS RELEVANCE TO TOXICOLOGICAL RESEARCH

Recent rapid advances in the science of genetics should provide a major stimulus to pharmacogenetics. Many loci associated with genetic diseases, including some associated with adverse drug reactions, have now been cloned and sequenced, and many more will be during the next few years. The development of the polymerase chain reaction makes it practical to genotype people using very small quantities of DNA from blood or even cheek scrapings. Eventually this will make it possible to take account of individual sensitivity when treating people with potentially toxic drugs, or when there is potential workplace exposure.

Until recently it was extremely difficult, and often impossible, to identify the genes which contribute to strain differences in response to xenobiotics. The development of sets of recombinant inbred strains provided a powerful tool in cases where there was only a single gene involved, but these crosses did not have the resolution to dissect out complex traits, involving two or more genetic loci (see, e.g., DiGiovanni et al., 1992). The only other method available was to study candidate genes. While this was quite successful in identifying loci concerned with drug metabolism, it was not able to detect novel genes such as those governing tissue sensitivity to a chemical. However, the recent development of microsatellite markers has completely revolutionized the genetic analysis of such complex polygenic characters. Microsatellites are short repetitive sequences of DNA with unique flanking regions which can be used as primers for the polymerase chain reaction (Dietrich et al., 1992; Love et al., 1990). Many of these sequences are polymorphic in the number of repeats, and it has been estimated that there are several thousand of these sequences scattered throughout the genome. More than 6000 have already been identified and mapped in the mouse, and information is available on their map location and primer sequence by e-mail (Copeland et al., 1993). Increasing numbers have been identified in the rat (Kunieda et al., 1992).

The basic technique involves identifying a pair of inbred strains which differ for the response of interest. These are then crossed to produce an F1 hybrid, and these are either backcrossed to the parental strains or intercrossed to produce F2 hybrids, or all three sets of crosses may be produced. The animals are then treated with the test compound and scored for sensitivity. This may be a discrete response (alive/dead), but, preferably, would involve a quantitative end point, such as the number of tumors. DNA is extracted from each individual, which is then typed for microsatellite markers for which the parental strains differ, covering all chromosomes. The aim is to identify an association between susceptibility and one or more of the markers. Such an association implies genetic linkage between the microsatellite marker and a susceptibility gene. Once this has been found, the location of the susceptibility gene is narrowed down using more markers on the same chromosome. Eventually there may be candidate genes close to the marker which would be worth considering as a susceptibility gene. If none can be found, then the susceptibility is probably governed by a new gene, which will need to be identified and cloned using the full range of molecular genetic techniques. Whether it would be worthwhile pursuing the gene further will depend on individual circumstances. Using these methods, Festing et al. (1994) were able to identify four genetic loci and the sex of the mouse as factors determining susceptibility to urethane-induced lung adenomas in mice.

VI. BETTER EXPERIMENTAL DESIGN COULD ACCOMMODATE MORE THAN ONE STRAIN AT LITTLE EXTRA COST

One of the problems to be overcome when encouraging research workers to use more than one strain is the fear that the size of the experiment will increase. This is not usually true. Experiments can be designed to study toxicity with two or more strains, which are no larger than would be needed in any case. This is done by using a "factorial" experimental design, and by ensuring that the experiment is no larger than necessary. This is discussed in more detail below. The essential requirement is to design "good" experiments.

Cox (1958) outlined the five requirements for a good experimental design. He stated that experiments should:

1. Be unbiased
2. Have high precision
3. Have a wide range of applicability
4. Be simple
5. Have the possibility of quantifying uncertainty

A full discussion of these requirements is beyond the scope of this chapter, but the need for high precision and the associated question of the size of the experiment and the need for a wide range of applicability are relevant, and will be discussed in more detail.

A. HIGH PRECISION IN AN EXPERIMENT
High precision in an experiment should ensure that if there is an effect due to the treatment, then there is a high probability that it will be detected. The way in which precision can be increased for quantitative end points is illustrated by reference to the formula for Student's t-test:

$$t = \frac{X_1 - X_2}{\sqrt{\dfrac{S_1^2}{n_1} + \dfrac{S_2^2}{n_2}}}$$

where t is Student's t, X_1 and X_2 are the means of the two groups, S_1^2 and S_2^2 are the within-group variances, and n_1 and n_2 are the numbers in the two groups.

High precision implies a high value of t. This can be achieved either by increasing the numerator, or by decreasing the denominator.

The size of the treatment effect (i.e., the difference between the two means in the numerator of the equation) can be maximized by using a higher dose level, the usual strategy used in toxicology, or by choosing more sensitive subjects which give a greater response, if such subjects are known. In some cases, the use of more than one strain will increase this sensitivity, though in experiments involving more than one strain and two or more treatments the data must be analyzed by two-way analysis of variance rather than Student's t-test.

The denominator can be reduced either by reducing the within-group variation, or by increasing the sample size. The within-group variation can be reduced by standardizing conditions, by careful selection of animals which are as uniform as possible, by using high-quality disease-free "SPF" animals, and by eliminating genetic variation by using isogenic (inbred or F1 hybrid) animals. These steps can very substantially increase the precision of an experiment (Festing, 1992). Some variation which cannot be eliminated even by these methods can be reduced by using a randomized block experimental design. Such a design breaks a large experiment down into smaller units, which are intrinsically more uniform. Randomized block designs are widely used in many disciplines, but this method of increasing precision seems to be neglected in toxicological research (Festing, 1992).

Finally, precision can be influenced by the size of the experiment, though most statistical textbooks give little guidance on choosing an appropriate size of an experiment. There are mathematical methods of estimating how large an experiment needs to be (Cohen, 1969), but these are usually impractical to apply to experiments with more than two treatments and with several dependent variables (characters). Most statisticians therefore use arbitrary rules, which are, however, easily justified. Mead (1988) suggested that the size of an experiment with a quantitative end point should be based on what he called the "resource equation":

$$(N - 1) = (T - 1) + (B - 1) + (E - 1)$$

where N, T, B, and E are the total number of observations, the number of treatments, the number of blocks, and the remainder (obtained by subtraction), respectively. He suggested that (E–1), which is called the "error degrees of freedom" should lie between about 10 and 20. With less than 10, the experiment

TABLE 9.2
Critical Values of Student's t at p = 0.05
for Various Degrees of Freedom

Degrees of freedom	Critical value of Student's t at p = 0.05
1	12.71
2	4.3
5	2.57
10	2.23
20	2.09
40	2.02
100	1.98

will be too small and will lack statistical power. However, increasing the size of the experiment so that E–1 is greater than about 20 is hardly worthwhile, because it only results in a marginal increase in precision. The reasons for this are shown in Table 9.2. This shows the 5% critical values of Student's t for various numbers of degrees of freedom. A calculated value of t must exceed this value for the treatment effect to be judged "significant" at the 5% probability level. Student's t is the ratio of the difference between the two treated groups divided by the standard error of this difference. With 1 degree of freedom, this ratio has to be greater than 12.71 for the difference to be judged "statistically significant" at the 5% level of probability. In other words, the difference between the treated and control group has to be very large if the experiment is very small. A larger experiment involving 12 observations (say 6 treated and 6 control) will have 10 degrees of freedom, and the ratio will only need to be 2.23 to be judged significant. If the size of the experiment is increased to 22 observations (20 degrees of freedom), then the ratio falls from 2.23 to 2.09. Approximately doubling the size of the experiment to 42 observations (40 degrees of freedom) only reduces the ratio from 2.09 to 2.02, which represents a quite trivial reduction of the size of treatment effect which will be judged to be significant.

The conclusion is that there are substantial benefits in increasing the total size of an experiment from 1 to 10 degrees of freedom for error, but increasing the size much beyond that which will give 20 degrees of freedom for error rapidly leads to diminishing returns. This has been illustrated for an experiment with just two treatments, but is equally true for more complex experiments involving several treatments.

B. INCREASING THE RANGE OF APPLICABILITY OF AN EXPERIMENT

If the outcome of every experiment depended critically on the exact conditions under which the experiment was conducted (e.g., temperature, bedding, type of cage, barometric pressure, light levels and colors, dietary composition, strain and species of animals, etc.), then it would never be possible to base decisions on the results of experiments. Usually, the outcome of a toxicological experiment does not depend to any significant extent on, say, the type of cage used or the temperature of the animal house. On the other hand, there are situations where temperature may be critical. The results of a toxicological experiment carried out in a tropical climate without air conditioning, and with the temperature always above 30°C may not be the same as the same experiment carried out at 20°C. Similarly, an experiment done with SD rats may not give the same results as one done with F344 rats.

The research worker is then faced with a dilemma. If the experiment is done with SD rats, will the results only apply to SD rats? If it is not safe to extrapolate from SD rats to F344 rats, is it safe to extrapolate to humans? Although a dilemma of this sort can never be fully resolved, it is at least possible to explore the effects of some of the variables which the research worker considers may be of critical importance. This can be done by designing factorial experiments which include more than one strain, diet, or whatever other variable the scientist considers might be important. This can usually be done without any substantial increase in the total size of the experiment, which would still be based on having 10 to 20 degrees of freedom for error, as outlined above.

A specific example may clarify this. Suppose an experiment is to be set up to study the effects of a control and two dose levels of a chemical on, say, hematological parameters. With three treatment groups, an experiment with 18 animals (6 per treatment group) would have $(N-1) - (T-1) = (18-1) - (3-1) = 15$ degrees of freedom for error (assuming no blocking), which falls in the acceptable range of between 10 and 20. This experiment would be analyzed using a one-way analysis of variance, followed by a range test to compare the three groups. Suppose that the research worker would like to see if the same results were obtained with both F344 and SD rats. Instead of doing the whole experiment twice (36 animals), it would be possible to use 3 SD and 3 F344 rats in each dose group. This would still involve 18 rats total. Now the degrees of freedom for error would be $(18-1) - (6-1) = 12$. This is only a small reduction in the number of degrees of freedom (from 15 to 12), so the precision of the experiment would not be reduced very much. Use of 24 animals in total would give $(24-1) - (6-1) = 18$ degrees of freedom for error, which would in fact increase precision. The experiment would now be analyzed by a two-way analysis of variance, with an estimation of the differences between the chemical treatment groups, an estimate of the strain differences in the hematological parameters, and an estimate of the interaction between the two, i.e., whether the strains are behaving in the same way in their response to the chemical. If the strains are behaving in the same way, then the research worker should feel safer in extrapolating to the species as a whole. If the strains respond differently, then much greater care would be necessary in the interpretation of the results.

More details on the design of factorial experiments are given by Mead (1988) and in most statistics textbooks.

VII. CONCLUSION

The thesis of this chapter is that there is often important genetic variation among both humans and laboratory animals in response to toxic chemicals. The type of variation which can be detected most easily differs, due largely to differences between humans and laboratory animals in population structure. In humans there are very large breeding populations in which the Mendelian segregation of genes governing sensitivity can be detected. Polygenic variation can be detected using twins, but twin research is difficult, and cannot be done ethically with highly toxic or carcinogenic chemicals. In contrast, with laboratory animals differences between strains are most easily detected, and these are very often polygenic. Although there are animal homologues of some human Mendelian variation, these are usually only detected if a specific effort is made to find them by sampling several different populations, and possibly even wild representatives of the species.

Most toxicological research and screening fails to take this genetic variation into account, even though experimental designs are available which could incorporate several strains without increasing the total size of the experiment. In the case of screening experiments, instead of using 50 genetically identical animals in each dose group, it would be possible to use 25 animals of each of 2 genotypes, or 10 animals of each of 5 genotypes, or any other combination to make up the 50 animals. Of course, the genotypes would need to be balanced across treatment groups. It would not be acceptable for the treated group to have more F344 rats than the control group because, for example, F344 rats develop a high incidence of testicular tumors. If the groups were unbalanced and the strain identity of each rat were not known, then it might be assumed that these testicular tumors were the result of the test chemical. Such balance is easy to arrange, by randomizing the strains to each treatment group separately and keeping a record of the genetic identity of each individual. If there were large strain differences in response, then this sort of design would be statistically more powerful than that based on a single strain, and in the absence of strain differences it would not be any less powerful.

In research, failure to take account of genetic factors has several unfortunate consequences. First, the generality of the results is never explored. The observed effect may be general across many members of the species, or it may be a unique effect found only in one particular strain. Second, strain differences may be a useful research tool for studying mechanisms of toxicity. Strain differences have been used successfully in this way for many years, but only by a minority of toxicologists. There is scope for this to become a more widely used approach, particularly now that genetic techniques are advancing so rapidly. Third, there may be human polygenic genetic variation homologous to that found between strains of laboratory animals. If the genes can be identified, cloned, and sequenced in animals, it should be possible to probe human genomic libraries to see if there are human homologues, and if there is polymorphism in humans comparable to that in the animals. Finally, experiments incorporating more than one strain can easily be designed and should not be appreciably larger than those using a single strain. The factorial experimental design is a very powerful design which provides a substantial scientific return for a very small investment in terms of increased resources and complexity. Some statisticians even suggest that all experiments should be designed in this way. To do otherwise is a waste of scientific resources.

REFERENCES

Anver, M. R., Cohen, B. J., Lattuda, C. P., and Foster, S. J., Age-associated lesions in barrier-reared male Sprague-Dawley rats: A comparison between Hap:(SD) and Crl:COBS[R]CD[R](SD) stocks, *Exper. Aging Res.,* 8, 3, 1982.

Bailey, D. W., Recombinant inbred strains, an aid to finding identity, linkage, and function of histocompatibility and other genes, *Transplantation,* 11, 325, 1971.

Biondi, G., Rickards, O., Guglielmino, C. R., and De Stefano, G. F., Marriage distances among the Afroamericans of Bluefields, Nicaragua, *J. Biosoc. Sci.,* 25, 523, 1993.

Bochkov, N. P. and Titenko, N. V., Genetic predisposition to environmental toxic agents: detection by distribution of ecogenetic markers, in *Ecogenetics: Genetic Predisposition to the Toxic Effects of Chemicals,* Grandjean, P., Ed., Chapman and Hall, London, 1991, 657.

Bonhomme, F. and Guenet, J.-L., The wild mouse and its relatives, in *Genetic Variants and Strains of the Laboratory Mouse,* 2nd Ed., Lyon , M. F. and Searle, A. G., Eds., Oxford University Press, 1989, 649.

Calabrese, E. J., Genetic predisposition to occupationally-related diseases: current status and future directions, in *Ecogenetics: Genetic Predisposition to the Toxic Effects of Chemicals,* Grandjean, P., Ed., Chapman and Hall, London, 1991, 19.

Chisholm, J. J., The continuing hazard of lead exposure and its effects on children, *Neurotoxicology,* 5, 23, 1984.

Cohen, J., *Statistical Power Analysis for the Behavioral Sciences,* Academic Press, New York, 1969.

Coleman, D. A., Trends in fertility and intermarriage among immigrant populations in Western Europe as measures of integration, *J. Biosoc. Sci.,* 26, 107, 1994.

Copeland, N. G., Jenkins, N. A., Gilbert, D. J., Eppig, J. T., Maltais, L. J., Miller, J. C., Dietrich, W. F., Weaver, A., Lincoln, S. E., Steen, R. G., Stein, L. D., Nadeau, J. H., and Lander, E. S., A genetic linkage map of the mouse: current applications and future prospects, *Science,* 262, 57, 1993.

Cox, D. R., *Planning Experiments,* John Wiley & Sons, New York, 1958.

Dietrich, W., Katz, H., Lincoln, S. E., Shin, H.-S., Friedman, J., Dracopoli, N. C., and Lander, E., A genetic map of the mouse suitable for typing intraspecific crosses, *Genetics,* 131, 423, 1992.

DiGiovanni, J., Imamoto, A., Naito, M., Walker, S., Beltran, L., Chenicek, J., and Skow, L. Further genetic analysis of skin tumor promotor susceptibility using inbred and recombinant inbred mice, *Carcinogenesis,* 13, 525, 1992.

Falconer, D. S., *Introduction to Quantitative Genetics,* 2nd Ed., Longman, London, 1981.

Felton, R. P. and Gaylor, D. W., Multistrain experiments for screening toxic substances, *J. Toxicol. Envirl. Health,* 26, 399, 1989.

Ferris, S. D., Sage, R. D., and Wilson, A. C., Evidence from mtDNA, sequences that common strains of inbred mice are descended from a single female, *Nature,* 295, 163, 1982.

Festing, M. F. W., The genetic reliability of commercially bred laboratory mice, *Lab. Anim.,* 8, 265, 1974.

Festing, M. F. W., A case for using inbred strains of laboratory animals in evaluating the safety of drugs, *Food Cosmet. Toxicol.,* 13, 369, 1975.

Festing, M. F. W., Properties of inbred strains and outbred stocks with special reference to toxicity testing, *J. Toxic. Envirtl. Health,* 5, 53, 1979a.

Festing, M. F. W., *Inbred Strains in Biomedical Research,* Macmillan Press, London, 1979b, 483.

Festing, M. F. W., The choice of animals in toxicological screening: inbred strains and the factorial design of experiment, *Acta. Zool. Pathol. Antvp.,* 75, 117, 1980a.

Festing, M. F. W., Inbred strains and the factorial experimental design in toxicological screening, in *Proc. 7th ICLAS Symp.,* Utrecht, Gustav Fischer, Stuttgart, 1980b, 59.

Festing, M. F. W., The case for isogenic strains in toxicological screening, *Arch. Toxicol. Suppl.,* 9, 127, 1986.

Festing, M. F. W., Genetic factors in toxicology: implications for toxicological screening, *CRC Crit. Rev. Toxicol.,* 18, 1, 1987.

Festing, M. F. W., Contemporary issues in toxicology: use of genetically heterogeneous rats and mice in toxicological research: a personal perspective, *Toxic. Appl. Pharmacol.,* 102, 197, 1990.

Festing, M. F. W., Genetic factors in neurotoxicology and neuropharmacology: a critical evaluation of the use of genetics as a research tool, *Experientia,* 47, 990, 1991.

Festing, M. F. W., The scope for improving the design of laboratory animal experiments, *Lab. Anim.,* 26, 256, 1992.

Festing, M. F. W., Genetic Variation in Outbred Rats and Mice and Its Implications for Toxicological Screening, *J. Exp. Anim. Sci.,* 35, 210, 1993a.

Festing, M. F. W., Revised list of inbred strains of mice, *Mouse Genome,* 93, 393, 1993b.

Festing, M. F. W. and Lovell, D. P., Domestication and development of the mouse as a laboratory animal, in *The Biology of the Laboratory Mouse,* Berry, R. J., Ed., *Symp. Zool. Soc. Lond.,* 47, 43, 1981.

Festing, M. F. W., Yang, A., and Malkinson, A. M., At least four genes and sex are associated with susceptibility to urethane-induced pulmonary adenomas in mice, *Genet. Res.,* 64, 99, 1994.

Gordon, C. J. and MacPhail, R. C., Strain comparisons of DFP neurotoxicity in rats, *J. Toxi. Envirtl. Health,* 38, 257, 1993.

Grandjean, P., Ed., *Ecogenetics: Genetic Predisposition to the Toxic Effects of Chemicals,* Chapman and Hall, London, 1991.

Green, M. C., Catalog of mutant genes and polymorphic loci, in *Genetic variants and strains of the Laboratory Mouse,* 2nd Ed., Lyon, M. F. and Searle, A. G., Eds., Oxford University Press, 1989, 12.

Greenhouse, D. D., Festing, M. F. W., Hasan, S., and Cohen, A. L., Catalog of inbred strains of rats, in *Genetic Monitoring of Inbred Rat Strains,* Hedrich, H., Ed., Gustav Fischer, Stuttgart, 1990, 410.

Hammond, P. B., Bornschein, R. L., and Zenick, H., Toxicological considerations in the assessment of lead exposure, *Neurotoxicology,* 5, 53, 1984.

Hanigan, M. H., Kemp, C. J., Ginsler, J. J., and Drinkwater, N. D., Rapid growth of preneoplastic lesions in hepatocarcinigen-sensitive C3H/HeJ male mice relative to C57BL/6J male mice, *Carcinogenesis,* 9, 885, 1988.

Haseman, J. K. and Hoel, D. G., Statistical design of toxicity assays: role of genetic structure of test animal population, *J. Toxic. Envirl. Health,* 5, 89, 1979.

Idle, J. R., Armstrong, M., Boddy, A. V., Boustead, C., Cholerton, S., Cooper, J., Daly, A. K., Ellis, J., Gregory, W., Hadidi, H., Hofer, C., Holt, J., Leathart, J., McCracken, N., Monkman, S. C., Painter, J. E., Taber, H., Walker, D., and Yule, M., The pharmacogenetics of chemical carcinogenesis, *Pharmacogenetics,* 2, 246, 1992.

Isaacs, J. T., Inheritance of a genetic factor from the Copenhagen rat and the suppression of chemically induced mammary adenocarcinogenesis, *Cancer Res.,* 48, 2204, 1988.

Kalow, W., *Pharmacogenetics,* W. B. Saunders, Philadelphia, 1962.

Kalow, W., Lessons from pharmacogenetics, in *Ecogenetics: Genetic Predisposition to the Toxic Effects of Chemicals,* Grandjean, P., Ed., Chapman and Hall, London, 1991, 125.

Kawano, S., Nakao, T., and Hiraga, K., Strain differences in butylated hydroxytoluene-induced deaths in male mice, *Toxicol. Appl. Pharmacol.,* 61, 475, 1981.

Kihara, M., Kihara, M., and Noda, K., Lung cancer risk of GSTM1 null genotype is dependent on the extent of tobacco smoke exposure, *Carcinogenesis,* 15, 415, 1994.

Kleeberger, S. R. and Levitt, R. C., Animal models for studies on genetic predisposition to adverse effects of chemical exposure, in *Ecogenetics: Genetic Predisposition to the Toxic Effects of Chemicals,* Grandjean, P., Ed., Chapman and Hall, London, 1991, 91.

Klein, J., Congenic and segregating inbred strains, in *Genetic Variants and Strains of the Laboratory Mouse,* 2nd Ed., Lyon , M. F. and Searle, A. G., Eds., Oxford University Press, 1989, 797.

Kunieda, T., Kobayashi, E., Tachibana, M., Ikadai, H., and Imamichi, T., Polymorphic microsatellite loci of the rat (Rattus norvegicus), *Mammalian Genome,* 3, 564, 1992.

Levy, G. N., Martell, K. J., DeLeon, J. H., and Weber, W., Metabolic, molecular genetic and toxicological aspects of the acetylation polymorphism in inbred mice, *Pharmacogenetics,* 2, 197, 1992.

Lindsey, R. J., Historical foundations, in Baker, H. J., Lindsey, J. R., and Weisbroth, S. H., Eds., *The Laboratory Rat,* Vol. 1, *Biology and Diseases,* New York, Academic Press, 1979, 1.

Love, J. M., Knight, A. M., McAleer, M. A., and Todd, J. A., Towards construction of a high resolution map of the mouse genome using PCR-analysed microsatellites, *Nucleic Acids Res.,* 18, 4123, 1990.

Lovell, D. P. and Festing, M. F. W., Relationships among colonies of the laboratory rat, *J. Hered.,* 73, 81, 1982.

Mailman, R. B. and Lewis, M. H., Neurotoxicants and central catecholamine systems, *Neurotoxicology,* 8, 123, 1987.

Malkinson, A. M., Nesbitt, M. N., and Skamene, E., Susceptibility to urethane-induced pulmonary adenomas between A/J and C57BL/6J mice: use of AXB and BXA recombinant inbred lines indicating a three-locus genetic model, *J. Natl. Cancer Inst.,* 75, 971, 1985.

McAuslane, J. A. N., Lumley, C. E., and Walker, R. S., The need for control animal pathology database: an international survey, *Human Exper. Toxicol.,* 10, 205, 1991.

Mead, R., *The Design of Experiments*, Cambridge University Press, 1988.

Nebert, D. W., Human genetic variation in enzyme detoxication in *Enzymatic Basis of Detoxication,* Vol. 1, Jakoby, W. B., Ed., New York, Academic Press, 1980a, 25.

Nebert, D. W., Pharmacogenetics: an approach to understanding chemical and biological aspects of cancer, *J. Natl. Cancer Inst.,* 64, 1279, 1980b.

Newton, J. F., Pasino, D. A., and Hook, J. B., Acetaminophen nephrotoxicity in the rat: quantitation of renal metabolic activation in vivo, *Toxical. Appl. Pharmacol.,* 78, 39, 1985.

Nishioka, Y., Y-chromosomal DNA polymorphism in mouse inbred strains, *Genet. Res. Camb.,* 50, 69, 1987.

Phillips, T. J., Feller, D. J., and Crabbe, J. C., Selected mouse lines, alcohol and behavior, *Experientia,* 45, 805, 1989.

Pohjanvirta, R., Juvonen, R., Karenlampi, S., Raunio, H., and Tuomisto, J., Hepatic Ah-receptor levels and the effect of 2,3,7,8-tetrachlorodibenzo-p-dioxin TCDD on hepatic microsomal monooxygenase activities in a TCDD-susceptible and -resistant rat strain, *Toxi. Appl. Pharmacol.,* 92, 131, 1988.

Rice, C. M. and O'Brien, S. J., Genetic variance of laboratory outbred Swiss mice, *Nature,* 283, 157, 1980.

Rodriguez-Larralde, A., Pavesi, A., Scapoli, C., Conterio, F., Siri, G., and Barrai, I., Isonomy and the genetic structure of Sicily, *J. Biosoc. Sci.,* 26, 9, 1994.

Shami, S. A., Grant, J. C., and Bittles, A. H., Consanguineous marriage within social/occupational class boundaries in Pakistan, *J. Biosoc. Sci.,* 26, 91, 1994.

Shellabarger, C. J., Stone, J. P., and Holtzman, S., Rat differences in mammary tumor induction with estrogen and neutron irradiation, *J. Natl. Cancer Inst.,* 61, 1505, 1978.

Shirai, T., Nakamura, A., Fukushima, S., Yamamoto, A., Tada, M., and Ito, N., Different carcinogenic responses in a variety of organs, including the prostate, of five different rat strains given 3,2′-dimethyl-4-aminobiphenyl, *Carcinogenesis,* 11, 793, 1990.

Slaga, T. J., Overview of tumour promotion in animals, *Environ. Health Perspectives,* 50, 3, 1983.

Smolen, T. N., and Smolen, A., Developmental expression of cocaine hepatoxicity in the mouse, *Pharmacol. Biochem. Behav.,* 36, 333, 1990.

Taylor, B. A., Recombinant inbred strains, in *Genetic Variants and Strains of the Laboratory Mouse,* 2nd Ed., Lyon, M. F. and Searle, A. G., Eds., Oxford University Press, 1989, 773.

van Zwieten, M. J., Shellabarger, C. J., Hollander, C. F., Cramer, D. V., Stone, J. P., Holtzman, S., and Broerse, J. J., Differences in the DMBA-induced mammary neoplastic response in two lines of Sprague-Dawley rats, *Eur. J. Cancer Clin. Oncol.,* 20, 1199, 1984.

Vesell, E. S. and Page, J. G., Genetic control of dicumarol levels in man, *J. Clin. Invest.,* 47, 2657, 1968a.

Vesell, E. S. and Page, J. G., Genetic control of drug levels in man: phenylbutazone, *Science,* 159, 1479, 1968b.

Vesell, E. S. and Page, J. G., Genetic control of drug levels in man: antipyrine, *Science,* 161, 72, 1968c.

Vulliamy, T., Mason, P., and Luzzatto, L., The molecular basis of glucose-6-phosphate dehydrogenase deficiency, *Trends in Genetics,* 8, 138, 1992.

Weir, D. M., Ed., *Handbook of Experimental Immunology 3: Genetics and Molecular Immunology,* Blackwell Scientific, Oxford, 1986.

Yamada, J., Nikaido, H., and Matsumoto, S., Genetic variability within and between outbred Wistar strains of rats, *Exper. Anim.,* 28, 259, 1979.

Chapter 10

MECHANISMS OF CELL DEATH, WITH PARTICULAR REFERENCE TO APOPTOSIS

Gerald M. Cohen, Marion MacFarlane, Howard O. Fearnhead,
Xiao-Ming Sun, and David Dinsdale

CONTENTS

I. Introduction 185
   A. Occurrence of Apoptosis 187
   B. Biochemical Characteristics of Apoptosis 187
   C. Genes Involved in Programmed Cell Death 188
   D. Agents Inducing Apoptosis 189

II. Degradation of DNA 189
   A. Dissociation of Internucleosomal Cleavage
      from Apoptosis 190
   B. Formation of Large Fragments of DNA
      Precedes Internucleosomal Cleavage 193

III. Preapoptotic Thymocytes 194
   A. Isolation of a Preapoptotic Population of Thymocytes 194
   B. Evidence that F3 Contains an Intermediate
      Population of Preapoptotic Thymocytes 197

IV. Model of Thymocyte Apoptosis 199

V. Apoptosis and Disease 199
   A. Apoptosis and Cancer 199
   B. Apoptosis and AIDS 202

References 202

I. INTRODUCTION

Cell death is an irreversible loss of structure and function. Cells die primarily by one of two major mechanisms, that is, by necrosis or apoptosis (Arends and Wyllie, 1991; Walker et al., 1988; Wyllie et al., 1980). Cell death by necrosis occurs as a result of a marked toxic or physical insult. For many years, cell death had been synonymous with necrosis and felt to be a result of massive

0-8493-9229-2/95/$0.00+$.50
© 1995 by CRC Press Inc.

tissue damage and widespread cell injury, often following exposure to high concentrations of chemicals. In necrosis, there occurs an irreversible increase in the permeability of mitochondrial and plasma membranes, resulting in the cells being stained by vital dyes such as trypan blue. *In vivo,* necrosis often affects groups of cells, and an inflammatory reaction is commonly observed. The histological appearance of necrosis is well documented, and there is good agreement of its ultrastructural features in a wide range of cells. At an early stage, when the mitochondria generally swell, autophagocytosis may increase. Mitochondrial matrix densities are often observed in necrotic cells committed to die. This is often accompanied by irregular dense clumping of chromatin, progressive dissolution of ribosomes, and focal disruption of plasma and organelle membranes. At later stages of necrosis, karyolysis is observed.

More recently there has been an upsurge of interest in another form of cell death termed apoptosis (Arends and Wyllie, 1991; Cohen et al., 1992a; Gerschenson and Rotello, 1992; Williams and Smith, 1993). Apoptosis is derived from the Greek word meaning "the falling of leaves in autumn" (Kerr et al., 1972). Apoptosis is a process in which individual cells die in a controlled manner in response to specific stimuli following an intrinsic program (Arends and Wyllie, 1991). It was originally defined as a controlled cell deletion, which appeared to play a complementary but opposite role to mitosis in the regulation of cell populations *in vivo* (Kerr et al., 1972) Apoptosis characteristically affects single cells and is not accompanied by an inflammatory reaction. Morphological changes during apoptosis have been well described and are characterized by condensation of chromatin, nucleolar disintegration, decrease in cell volume, and dilatation of the endoplasmic reticulum (Wyllie et al., 1980; Walker et al., 1988). Volume reduction is such a prominent feature of apoptosis that initially this type of cell death was known as shrinkage necrosis (Kerr, 1971). Following the early morphological changes, fragmentation of the cell into apoptotic bodies often occurs. These are contained within intact membranes and thus apoptotic cells and apoptotic bodies, at least at early times, are not stained with vital dyes. On prolonged incubation *in vitro,* apoptotic cells lose their membrane integrity and thus include vital dyes. This is often referred to as secondary necrosis.

The biochemical hallmark of apoptosis has been considered to be the internucleosomal cleavage of DNA into nucleosomal fragments of 180 to 200 base pairs or multiples thereof (Wyllie, 1980). Apoptotic cells and bodies are rapidly phagocytosed in vivo, so that their half-life in organs such as the thymus and liver is very short: of the order of 2 to 4 hours (Bursch et al., 1990). The very short half-life of apoptotic cells *in vivo* and consequently the difficulty in recognizing and quantifying these cells has been a major reason why the importance of this major mechanism of cell death has, until recently, been relatively neglected. A comparison of some of the characteristics of apoptosis and necrosis is given in Table 10.1.

TABLE 10.1
A Comparison of Apoptosis and Necrosis

Apoptosis	Necrosis
Affects single cells	Affects groups of cells
Chromatin marginates as a large crescent	Ill-defined clumping of chromatin
Internucleosomal cleavage of DNA to give characteristic DNA ladder	Random cleavage
Formation of large fragments of DNA (30–50 kilobase pairs)	Random cleavage
Decrease in cell volume	Increase in cell volume
Organelles retain integrity	Swelling of organelles
Cell fragments into small apoptotic bodies	Cell swells and lyses
Apoptotic cells and bodies are phagocytosed without causing an inflammatory reaction	Cell contents are released triggering an inflammatory reaction

A. OCCURRENCE OF APOPTOSIS

Apoptosis plays a key role in the control of cellular populations in development, in the immune system, and in carcinogenesis (Arends and Wyllie, 1991;Cohen et al., 1992a; Williams and Smith, 1993). It is now recognized that most if not all physiological cell death occurs by apoptosis. This process is important in development — for example, cells required only in the male being are discarded in the female. In the developing nervous system, neurones that fail to make the correct connections die by apoptosis. Apoptosis has also been observed during development of the gut and in remodeling of cartilage and bone. In the immune system, apoptosis is also an important mechanism of cell death (Arends and Wyllie, 1991; Cohen et al., 1992a). Immature thymocytes are removed when they fail to rearrange their antigen receptors correctly, if they are self reactive or if they do not recognize foreign antigens.

B. BIOCHEMICAL CHARACTERISTICS OF APOPTOSIS

The most distinctive biochemical characteristic of apoptosis is the "DNA ladder" observed on agarose gel electrophoresis. This arises from the internucleosomal cleavage of DNA resulting in nucleosomal fragments of 180 to 200 base pairs and multiples thereof. This characteristic was first recognized by Wyllie (1980) in thymocytes treated with glucocorticoids and has since been observed in numerous other systems. The origin of the DNA ladder is believed to be the activation of a Ca^{2+}/Mg^{2+}-dependent endonuclease, which then cleaves DNA at internucleosomal regions. Despite extensive attempts in many laboratories, the endonuclease has not yet been purified. Some reports have suggested that the endonuclease may be NUC-18 (Gaido and Cidlowski, 1991), DNase I (Ucker et al., 1992; Peitsch et al., 1993), or DNase II (Barry and Eastman, 1992). Recently several laboratories, including our own, have shown a dissociation of internucleosomal DNA ladder formation from the morphological features of apoptosis (discussed below). Despite these observations, the recognition of a ladder is often a valuable and simple biochemical indicator of the likely occurrence of apoptosis. The presence of a DNA ladder is not proof *sine qua non* of apoptosis, but is highly indicative, particularly if morphological evidence is also obtained.

It has generally been accepted that apoptosis in thymocytes is dependent on new protein and RNA synthesis. This conclusion has been based on the inhibition of the induction of apoptosis in thymocytes, various T cells, and in T cell hybridomas by inhibitors of both protein synthesis, such as cycloheximide and emetine, and of RNA synthesis, such as actinomycin D (Wyllie et al., 1984; Cohen et al., 1992a). These observations have added weight to the suggestion that apoptosis in thymocytes is an active process and requires new protein synthesis by "death genes." Conversely, however, protein and RNA synthesis inhibitors actually induce apoptosis in several other systems, particularly human tumor cell lines (Martin, 1993; Cohen et al., 1992a). This mechanism is not fully understood, but it has been suggested that the inhibitors prevent the synthesis of rapidly turning over proteins, which are required to prevent the cells from entering apoptosis.

C. GENES INVOLVED IN PROGRAMMED CELL DEATH

Undoubtedly the best system for the study of the genetics of programmed cell death is the nematode *Caenorhabditis elegans*. During development, 131 cells of the 1090 cells of this worm are deleted by a process resembling apoptosis. To date 14 genes have been shown to be involved in various parts of the death program (Ellis et al., 1991). Two of these genes, *ced-3* and *ced-4,* must function for the cells to undergo programmed cell death. The *ced-4* gene encodes a novel protein, which has two sites similar to the Ca^{2+}-binding EF-hand motif (Ellis et al., 1991). The *ced-3* gene encodes a novel protein, 503 amino acids in length, containing a serine-rich middle region of about 100 amino acids (Yuan et al., 1993). Since serines are common phosphorylation sites, it is possible that the ced-3 protein and programmed cell death in *C. elegans* are controlled by phosphorylation (Yuan et al., 1993). The non-serine rich region appears to be related to interleukin-1β converting enzyme, a cysteine protease which cleaves the inactive 31 kd precursor of IL-1β to the active cytokine (Thornberry et al., 1992). Genes such as *ced-3* and *ced-4* encode cytotoxic moieties, which must generally be under very tight control in order to avoid damaging the wrong cells. The *ced-3* and *ced-4* genes are negatively regulated by the *ced-9* gene. Recent findings have shown significant sequence homology between *ced-9* and *bcl-2,* and mammalian *bcl-2* can protect cells in the nematode from death (Vaux, 1993).

The function of *bcl-2* in mammalian cells is not known. While it was originally thought to be mitochondrial in origin, more recent studies have shown it to have a more widespread cellular distribution, including the nucleus and endoplasmic reticulum. Induction of apoptosis by diverse stimuli including growth factor withdrawal, radiation, hyperthermia, glucocorticoids, and many different chemotherapeutic agents, is inhibited by *bcl-2* (Vaux et al., 1988; Hockenbery et al., 1990). Bcl-2 protects against many but not all, types of physiological cell death. For example *bcl-2* expression in immature thymocytes protects against apoptosis induced by anti-CD3 antibodies, gluco-corticoids, and radiation, but it does not prevent deletion of autoreactive T cells (Sentman et al., 1991; Strasser et al., 1991). Thus, in this example, there are at least two pathways of apoptosis, a *bcl-2* dependent and independent pathway. A recent report suggests that *bcl-2* may be of significance in protection against free radicals and oxidative stress (Hockenbery et al., 1993). It is of interest that in *C. elegans* the final step in the apoptotic process involves dissolution of the nucleus by a nuclease encoded by the *nuc-1* gene. In *nuc-1* mutants, cell death and engulfment still occur, but the pyknotic DNA of the dead cells is not degraded (Ellis et al., 1991). Which of the genes involved in cell death in the nematode have mammalian counterparts is of great interest.

Various other genes have been suggested to be involved in mammalian apoptosis including the early response genes, *c-fos* and *c-jun* as well as *c-myc, p-53, clusterin* (also known as *TRPM 2* or *SGP1*), *RP-2,* and *RP-8* (reviewed in Oren, 1992). The role(s) of these genes in apoptosis is currently the subject of intensive investigations in many laboratories.

D. AGENTS INDUCING APOPTOSIS

A wide variety of diverse agents have been shown to induce apoptosis in many different systems. One of the best characterized model systems is the induction of apoptosis in thymocytes. Numerous agents, including glucocorti-coids (Wyllie, 1980; Cohen and Duke, 1984), etoposide (Walker et al., 1991; Cohen et al., 1993), irradiation, tributyltin oxide (Aw et al., 1990; Raffray and Cohen, 1991), anti-CD3 antibodies (Smith et al., 1989), and thromboxane A2 agonists (Ushikubi et al., 1993) have been shown to induce apoptosis in thymocytes or in T cell derived lines. It is readily apparent that these include both physiological and pathological stimuli. Among the latter is a wide diver-sity of chemicals with a range of different chemical structures and biological activities. It has been suggested that cells, such as thymocytes, are primed to die and that the trigger can be activated by a number of different events (Arends and Wyllie, 1991).

II. DEGRADATION OF DNA

Thymocytes have been one of the most extensively studied model systems for investigating mechanisms of apoptosis. As a result of this, the biochemical and morphological features of apoptosis in these cells have had a major influence on our thinking in this field. As the nucleus is the dominant organelle in thymocytes, many studies have concentrated on morphological changes to the nucleus and to the degradation of DNA. Recent studies from our laboratory, as well as several others, have shed new light on the mechanism of apoptosis, particularly on the degradation of DNA in thymocytes. In the remainder of this article, we shall describe some of these advances.

A. DISSOCIATION OF INTERNUCLEOSOMAL CLEAVAGE FROM APOPTOSIS

Recently we have described a novel flow cytometric method to separate and quantify normal and apoptotic thymocytes (Sun et al., 1992). Thymocytes were isolated from immature rats and incubated with dexamethasone (0.1 μM) for periods up to 4 hours. Subsequent incubation with the vital bisbenzimidazole dye Hoechst 33342 and the DNA intercalating agent propidium iodide enabled three distinct populations of cells to be identified. Nonviable cells fluoresced red due to uptake of propidium iodide, whereas normal and apoptotic cells fluoresced blue due to Hoechst 33342. Apoptotic cells were distinguished from normal thymocytes both by their higher intensity of blue fluorescence and their smaller size as determined by a reduction in forward light scatter (Figure 10.1). Cells from these populations were purified by fluorescence activated cell sorting (FACS). The larger cells with low blue fluorescence showed normal thymocyte ultrastructure and the absence of any DNA fragmentation, as assessed by agarose gel electrophoresis. In contrast, the smaller cells showed both the morphological characteristics of apoptosis and extensive internucleosomal fragmentation of DNA to multiples of approximately 180 base pairs (Sun et al., 1992). Treatment of thymocytes with dexamethasone for 4 hours caused an induction of apoptosis as determined by the increase in the smaller cells with high blue fluorescence (compare Figures 10.1a and b). The induction of apoptosis was inhibited, as expected, by the protein synthesis inhibitor, cycloheximide (Figure 10.1c).

Zinc has also been reported to prevent apoptosis, an effect assumed to be due to the ability of the metal ion to inhibit a Ca^{2+}/Mg^{2+}-dependent endonuclease (Cohen and Duke, 1984; Waring et al., 1990). It was therefore surprising that zinc did not decrease the formation of apoptotic cells, as judged by the property of high blue fluorescence (region 1 in Figure 10.1d), under conditions where zinc almost completely inhibited both DNA fragmentation and DNA laddering induced by dexamethasone (Figure 10.2) (Cohen et al., 1992b). Ultrastructural examination of these cells following FACS showed that the thymocytes were shrunken, with dilation and "bubbling" of the smooth endoplasmic reticulum (Figure 10.3). The heterochromatin in these cells was condensed and arranged in several sharply defined clumps, which were contiguous with the nuclear membrane (Figure 10.3b). Similar nuclear changes have been described, in various cell types, as the earliest signs of apoptosis (Kerr et al., 1987; Walker et al., 1988). In the presence of zinc, many of the cytoplasmic changes characteristic of apoptosis were still evident. Thus the nuclear changes in the cells were probably arrested at an early stage of apoptosis prior to the effects of the endonuclease. In the presence of zinc, dexamethasone induces many key features of apoptosis without any evidence of DNA laddering. Similar results were observed in thymocytes treated with etoposide, a DNA topoisomerase II inhibitor, in the presence of zinc (Sun et al., 1994). These results suggest that critical changes occur during the induction of apoptosis, both in nuclear DNA and in the cytoplasm prior to endonuclease cleavage of DNA into nucleosomal fragments. While our results are consistent with a role for zinc in inhibiting the endonuclease (Cohen and Duke, 1984; Waring et al., 1990), they do not exclude other possibilities such as zinc binding to the DNA. Zinc may also modify apoptosis by activating protein kinase C or by inhibiting phosphorylases associated with inositol phosphate metabolism (Waring et al., 1991).

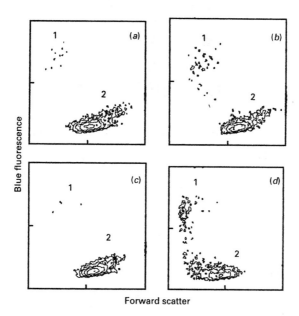

FIGURE 10.1. Zn^{2+} does not prevent the formation of apoptotic thymocytes by dexamethasone. Freshly isolated thymocytes were incubated for 4 hours either alone (a) or with dexamethasone (0.1 μM) in the absence (b) or in the presence of either cycloheximide (10 μM) (c) or zinc acetate dihydrate (1 mM) (d). Normal (region 2) and apoptotic (region 1) thymocytes were separated by flow cytometry following incubation with Hoechst 33342 (1 μg/ml) and propidium iodide. Non-viable cells, which included propidium iodide, were gated out. Apoptotic cells, with high blue fluorescence and low forward light scatter, were distinguished from normal thymocytes, with low blue fluorescence and high forward light scatter. Forward scatter gives an indication of size, i.e., apoptotic cells were smaller than normal cells. (Reproduced with permission from Cohen, G. M., Sun, X.-M., Snowden, R. T., Dinsdale, D., and Skilleter, D. N., Key morphological features of apoptosis may occur in the absence of internucleosomal DNA fragmentation, *Biochem. J.,* 286, 331, 1992b.)

STD   CON DEX DEX+Zn

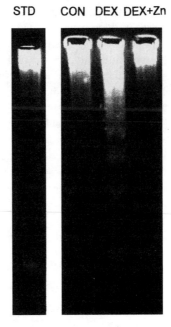

FIGURE 10.2. Zn^{2+} inhibits dexamethasone-induced DNA laddering in thymocytes. Thymocytes (20×10^6 cells) were incubated for 4 hours either alone (CON) or with dexamethasone (0.1 μM) in the absence (DEX) or in the presence of Zn^{2+} (1 mM) (DEX + Zn), and 1×10^6 cells were taken for measurement of DNA laddering by agarose gel electrophoresis. Lane 1 contains molecular size standards of multiples of 123 bp. In unsorted thymocytes, dexamethasone (lane 3) caused an increase in DNA laddering compared with control cells (lane 2). This increase was totally inhibited in the presence of Zn^{2+} (1 mM) (lane 4).

This, to our knowledge, was the first time in thymocytes that a clear dissociation had been observed between the morphological features of apoptosis and DNA laddering. A similar absence of DNA laddering has been observed in some cases of programmed cell death, for example in insect metamorphosis and in normal limb development (Lockshin and Zakeri, 1991). In addition, chromatin condensation characteristic of apoptosis has been observed in oligo-dendrocytes (Barres et al., 1992), hepatocytes (Oberhammer et al., 1992), and in murine embryonic fibroblasts (Tomei et al., 1993) again in the absence of DNA ladder formation. Our results with zinc support the hypothesis that the induction of the earliest morphological changes of apoptosis involve enzymes other than the Ca^{2+}/Mg^{2+}-dependent endonuclease and that this endonuclease is not involved until the later stages, when it is responsible for internucleosomal fragmentation of the DNA (Wyllie, 1980). This is analogous to the situation with *nuc-1* in *C. elegans,* discussed above, when the nuclease appears to be involved in the mopping up process after cell death has already occurred.

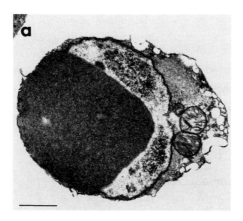

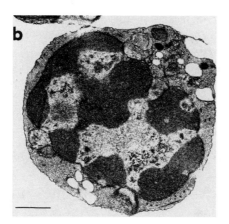

FIGURE 10.3. Zinc arrests some of the apoptotic changes in dexamethasone-treated thymocytes. Thymocytes treated with dexamethasone demonstrated the characteristic chromatin condensation and cytoplasmic contraction of apoptosis (a). The heterochromatin was condensed and usually coalesced against one pole of the nuclear membrane. Cells treated with dexamethasone in the presence of zinc showed only the earliest signs of apoptosis in the nucleus (b). Discrete clumps of condensed chromatin were abutted against the nuclear membrane; a further clump was present in the center of many of these nuclei. The nuclear membrane, although usually intact, was often convoluted. The euchromatin retained its normal density, but often included one or more clusters of intensely-stained nucleolar remnants. Bars represent 1 μm.

B. FORMATION OF LARGE FRAGMENTS OF DNA PRECEDES INTERNUCLEOSOMAL CLEAVAGE

In our initial investigations, we demonstrated that zinc inhibited internucleosomal cleavage of DNA, but did not inhibit many of the morphological features of apoptosis. As the cells treated with dexamethasone in the presence of zinc showed distinct nuclear changes (Figure 10.3b), we wished to investigate whether there were any other alterations in DNA. In order to accomplish this, we used field inversion gel electrophoresis (FIGE). This method allows one to determine the size of DNA fragments up to 450 kilobase pairs in length, in contrast to conventional agarose gel electrophoresis, which only allows analysis of DNA fragments up to approximately 20 kilobase pairs.

Using this method, we demonstrated that with dexamethasone alone large fragments of DNA, particularly of 30 to 50 kilobase pairs in length, were formed, and the formation of these fragments preceded internucleosomal cleavage of DNA (Brown et al., 1993). However, the results in the presence of zinc were particularly interesting. A marked accumulation of fragments of 30 to 50 kilobase pairs was observed as well as the appearance of fragments of 200 to 250 kilobase pairs (Figure 10.4). The appearance of these large fragments was coincident with the morphological changes observed in these cells. These results suggested that the formation of these large fragments was a critical event in glucocorticoid -induced apoptosis in thymocytes. As the endonuclease cleavage of DNA into nucleosomal ladders is mediated by a zinc inhibitable Ca^{2+}/Mg^{2+}-dependent enzyme, our data suggested that other key enzyme(s) were responsible for the formation of the large fragments of DNA. The significance of the sizes of the large fragments is not clear. Fragments of approximately 50 and 250 kilobase pairs may represent cleavage of chromosomal domains such as supercoiled loops and rosettes, respectively (Filipski et al., 1990; Walker et al., 1991).

III. PREAPOPTOTIC THYMOCYTES

A. ISOLATION OF A PREAPOPTOTIC POPULATION OF THYMOCYTES

In apoptotic thymocytes, the loss of cell volume in response to dexamethasone, irradiation, and etoposide has been associated with a single stepwise increase in buoyant density (Wyllie and Morris, 1982; Walker et al., 1991). In several studies of apoptosis, normal and apoptotic thymocytes have been isolated without any evidence of cells of an intermediate size (Wyllie and Morris, 1982; Walker et al., 1991) and thus by implication, no intermediate population (Cohen et al., 1992a). Recently we have identified and isolated such an intermediate population of cells. These cells are the immediate precursors of an apoptotic cell population and represent cells at an early stage of apoptosis. Using discontinuous Percoll density gradients, we obtained four fractions of cells (F1–F4) in order of increasing density and decreasing size. We have separated and characterized these fractions. Of particular interest was F3, which contained cells intermediate in size and density between the normal cells in F1 and F2 and the apoptotic cells in F4 (Cohen et al., 1993).

Thymocytes were incubated with dexamethasone (0.1 μM) or etoposide (10 μM) for 4 hours and then separated on Percoll and characterized by various techniques. Examination by flow cytometry, following incubation with Hoechst

33342, showed that cells in F1 and F2 exhibited high forward light scatter and low blue fluorescence, typical of normal cells (Figures 10.5a and b). Cells in F3 were heterogeneous, containing both smaller cells with high blue fluorescence (region 3, Figure 10.5c), as well as apparently normal cells (region 1, Figure 10.5c), which exhibited low blue fluorescence but were somewhat smaller (lower mean forward light scatter) than the corresponding cells from

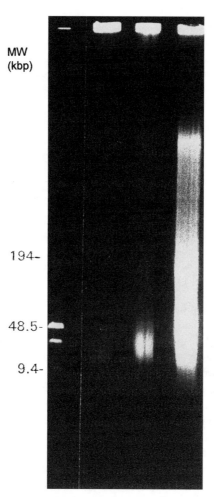

FIGURE 10.4.   Incubation of thymocytes with dexamethasone in the presence of zinc leads to the accumulation of large molecular weight fragments of DNA. Thymocytes were incubated for 4 hours either alone (CON), in the presence of dexamethasone (0.1 μM) either alone (DEX) or with zinc (1mM) (Zn). Lane 1 contains 0.1–200 kilobase pair standards.The numbers at the left hand side of the figure represent the size of the standards in kilobase pairs. Agarose plugs were prepared and separated by field inversion gel electrophoresis (Brown et al., 1993). In the presence of zinc, a clear accumulation of large fragments of DNA was observed.

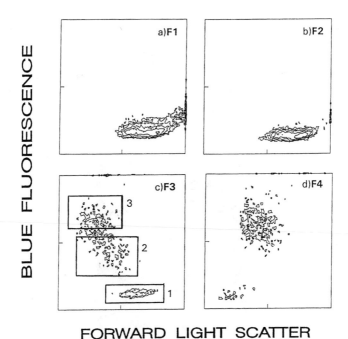

FORWARD LIGHT SCATTER

FIGURE 10.5.   Flow cytometric analysis of cells fractioned from Percoll. Thymocytes were incubated with etoposide (10 μM) for 4 hours. These cells were separated on Percoll and the four fractions of increasing density from F1 to F4 incubated with Hoechst 33342 at 37° for a further 10 min and then examined by flow cytometry. Reproduced with permission from Cohen et al., *J. Immunol.,* 151, 566, 1993. Copyright 1993, *The Journal of Immunology.*

F1 and F2. A third group of cells, present within F3 (region 2, Figure 10.5c) had properties intermediate between the other two populations. Cells in F4 were predominantly of a smaller size (lower forward light scatter) and exhibited high blue fluorescence with Hoechst 33342 (Figure 10.5d).

Agarose gel electrophoresis of DNA from cells fractionated following exposure to etoposide showed little or no DNA laddering in the normal cells in F1 and F2, whereas extensive laddering was observed in the purported apoptotic cells in F4. Significant laddering was also observed in cells from F3, but it was less than that observed from F4 (Cohen et al., 1993).

Cells from the four different Percoll fractions, prepared after treatment with etoposide for 4 hours, were also processed for electron microscopy. The morphology of the majority (>85%) of the cells in F1 and F2 was indistinguishable from that of most thymocytes isolated from untreated rats. The nuclei in both fractions were rounded and characterized by relatively condensed perinuclear heterochromatin, which lined the nuclear membrane except for small regions around the nuclear pores. This heterochromatin extended as several, ill-defined projections into the euchromatin. The cell population of highest buoyant density (F4) was primarily apoptotic, most of which had accumulations of condensed chromatin forming dense apical caps within the nucleus. Discontinuities in the nuclear membrane were often evident within the cytoplasm of these cells, and nucleolar remnants, particularly the dense fibrillar components, were prominent. The cisternae of the endoplasmic reticulum were often dilated, and many of the resulting vacuoles fused with the cell membrane, usually toward one pole of the cell (Cohen et al., 1993).

Unlike the other three fractions, F3 was strikingly heterogeneous. It contained a few apoptotic bodies (<5%) and many profiles (40 to 50%) exhibited the features of cells in F4. The remaining profiles were cells with a distinctly different morphology. These cells had a rather dense, granular cytoplasm containing normal organelles, except for a few small vacuoles or dilated cisternae of the endoplasmic reticulum. The most striking feature of these cells was the compaction of the perinuclear heterochromatin into dense, sharply-defined clumps abutting onto the nuclear membrane. Unusually large regions of this membrane were devoid of heterochromatin, and the membrane itself was often convoluted. The euchromatin and fibrillar centers of these cells were apparently normal.

B. EVIDENCE THAT F3 CONTAINS AN INTERMEDIATE POPULATION OF PREAPOPTOTIC THYMOCYTES

In order to test the hypothesis that the cells in F3 were an intermediate population between the normal (F1 and F2) and the apoptotic thymocytes (F4), thymocytes were incubated with etoposide (10 µM) for 4 hours, and following Percoll separation the cells in F3 were further sorted by flow cytometry into the population of cells of either low blue fluorescence (region 1, Figure 10.5c), or of intermediate and high blue fluorescence (regions 2 and 3, Figure 10.5c). On examination by agarose gel electrophoresis, no DNA laddering was observed with cells from region 1, whereas significant laddering was observed with cells from regions 2 and 3. Ultrastructural examination of the cells from region 1 by electron microscopy showed that this population was homogeneous (Figure 10.6) and characterized by distinctive perinuclear accumulations of condensed heterochromatin. The sharp demarcation between these clumps and the euchromatin was accentuated by the apparent loss of the perichromatin fibrils, which permeated this region in both untreated cells and in the etoposide-treated cells collected in F1 and F2. In marked contrast, most of the cells (>95%) in regions 2 and 3 showed the ultrastructural features associated with apoptotic cells (see Figure 10.3a). In addition, when a homogeneous population of cells from F3 was isolated, no DNA ladders were observed. On further incubation for 2 hours in the presence or absence of etoposide (10 µM), the cells became smaller, exhibited higher blue fluorescence with Hoechst 33342, and developed DNA ladders.

Thus, in these studies we have identified and partially characterized a transitional population of cells, intermediate in size between normal and apoptotic

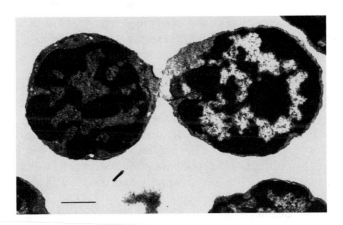

FIGURE 10.6. Ultrastructure of preapoptotic thymocytes. These cells showed aggregation of heterochromatin into discrete clumps, which were sharply delineated from the remaining nucleoplasm. Thymocytes were incubated for 4 hours in the presence of etoposide and separated into four fractions (F1–F4) by isopycnic Percoll centrifugation. At least two types of cell were evident in F3, which were subsequently sorted by flow cytometry according to their fluorescence intensity with Hoechst 33342. Cells with high blue fluorescence were apoptotic, similar to those in F4, whereas those with low blue fluorescence were identified as preapoptotic.

thymocytes, which exhibits a distinctive morphology. No internucleosomal cleavage of DNA was observed in these transitional cells until the dramatic changes of nuclear morphology, characteristic of apoptosis, were observed. This observation provides strong evidence for a causative link between these two events. Cells in F1 and F2 had normal morphology, while those in F4 were typically apoptotic. All these fractions were homogeneous as judged by both fluorescence characteristics with Hoechst 33342 and electron microscopy. In marked contrast, after 4 hours incubation with etoposide, cells in F3 were heterogeneous both by flow cytometry (Figure 10.5c) and electron microscopy (Cohen et al., 1993). Separation of these heterogeneous cells by FACS showed subtle, but distinct, changes in the distribution of heterochromatin in most cells from region 1 (Figure 10.6). These changes resemble those observed in thymocytes in which the process of dexamethasone-induced apoptosis has been arrested by the presence of zinc (Cohen et al., 1992b). This peripheral coalescence of heterochromatin has been recognized by several workers as an early sign of apoptosis (Kerr et al., 1987; Walker et al., 1988; Bursch et al., 1992). It probably indicates major changes in the DNA but, in the absence of any laddering in this fraction, it clearly precedes internucleosomal cleavage. In contrast, cells from regions 2 and 3 (Figure 10.5c) exhibited a dramatic change in morphology demonstrating all the characteristic changes associated with apoptosis, together with DNA laddering. These data provide strong evidence that internucleosomal cleavage of DNA is responsible for the most characteristic morphological changes in the nucleus of apoptotic thymocytes, in agreement with the suggestion of Arends et al. (1990).

Thus our data strongly suggest the formation of a preapoptotic population of thymocytes intermediate in size and density between normal and apoptotic cells. We have defined these cells as preapoptotic because they do not exhibit internucleosomal cleavage of DNA but are nevertheless committed to become apoptotic. The properties of these cells are summarized in Table 10.2. The isolation of such a population of cells will greatly facilitate mechanistic studies on apoptosis, in part because they enable one to establish whether the changes observed are an early or late stage in the process.

IV. MODEL OF THYMOCYTE APOPTOSIS

Based on work from our laboratory, as well as several others, it is possible to propose a model to explain the relationship between DNA degradation, chromatin structure, and changes in nuclear morphology. We propose that normal thymocytes in F1 and F2 (Figure 10.5) are converted to preapoptotic cells in F3, and this coincides with the formation of large fragments of DNA. These large fragments are possibly associated with the areas of condensed chromatin abutting the nuclear membrane in cells in F3. Normally these large fragments are rapidly cleaved by a zinc-inhibitable Ca^{2+}/Mg^{2+}-dependent endonuclease to give rise to characteristic apoptotic cells (F4) possessing DNA ladders. This model is depicted in Figure 10.7.

V. APOPTOSIS AND DISEASE

As physiological or programmed cell death is such an important biological process, alterations or defects in it may well result in pathological consequences. Defective regulation of apoptosis may play a role in a number of diseases including the induction and progression of cancer, AIDS, autoimmune diseases, and degenerative diseases of the central nervous system (Bursch et al., 1992; Carson and Ribeiro, 1993). In this short review, we shall briefly consider the importance of apoptosis in cancer and AIDS.

A. APOPTOSIS AND CANCER

Normal tissue homeostasis is maintained by a balance of cell input, due to cell proliferation and renewal, and cell death. Cancer represents an imbalance,

TABLE 10.2
Properties of Preapoptotic
Thymocytes

Intermediate in size and density
Decreased CD4 and CD8
No internucleosomal cleavage
50kbp fragments of DNA
Increased cell membrane permeability
Condensed chromatin abutting nuclear membrane

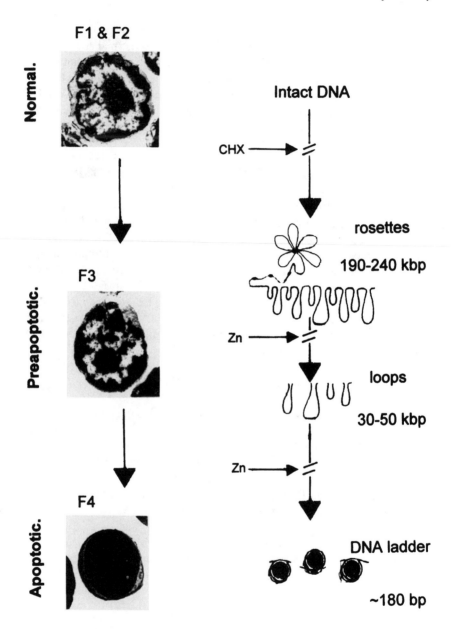

FIGURE 10.7.   Model of DNA degradation in thymocyte apoptosis. Intact DNA of normal thymocytes (F1 and F2) is initially cleaved into large fragments present at the earliest stages of apoptosis (F3) coincident with the condensation of the chromatin abutting the nuclear membrane. These cells are rapidly converted into cells with the classical ultrastructure and biochemistry of fully apoptotic cells (F4). Reproduced with permission from Cohen et al., *J. Immunol.,* 153, 507, 1994. Copyright 1994, *The Journal of Immunology.*

which may be due to an increase in cell proliferation and/or a decrease in cell death. The induction of cancer is a multistep process involving tumor initiation, promotion, and progression. Recent studies have demonstrated that apoptosis may intefere with tumor formation at one or more of these steps in the carcinogenic process. Many genotoxic carcinogens initiate cancer by an initial damage to DNA. This may result in the cell delaying cell division until the damage has been repaired, or the cell may undergo apoptosis, or it may continue through the next cell cycle. Apoptosis is a very efficient means of preventing cancer because it can remove cells with genetic lesions. A defect in the ability of a cell with damaged DNA to undergo apoptosis may therefore lead to an increase in cancer incidence. The factors which determine the fate of a cell with damaged DNA are not clearly known, but there seems to be a role for some oncogenes and tumor suppressor genes such as *c-myc, bcl-2,* and *p53.* The *c-myc*-proto-oncogene has been implicated in the control of normal cell growth and its deregulation in the development of neoplasia (Spencer and Groudine, 1991). More recent studies have shown an association of deregulated *c-myc* with an increased incidence of apoptosis, in particular under conditions of restrained cell proliferation (Evan et al., 1992), for example, *c-myc* accelerated apoptosis in myeloid cells, on withdrawal of interleukin-3, and in primary fibroblasts deprived of serum. Thus *c-myc* protein can induce either cell proliferation or cell death dependent on the conditions. In order to explain this apparent paradox, it has been suggested that cell proliferation and cell death pathways are both tightly coupled processes driven by *c-myc* (Evan and Littlewood, 1993). Once these pathways are established, they are then independently regulated by other genes, cytokines, and other external factors. The biological advantage of this model is that all proliferating cells, which may potentially be a carcinogenic hazard, are also primed to die if they do not receive appropriate survival signals.

In contrast to *myc,* over-expression of the *bcl-2* gene makes cells resistant to apoptosis without affecting proliferation, as discussed earlier. Follicular B-cell lymphomas have high concentrations of *bcl-2* protein, due to a t(14:18) translocation that places the *bcl-2* gene under the control of the strong immunoglobulin heavy chain gene promoter. Thus tumor cells possessing this translocation would have a survival advantage over normal cells (Korsmeyer, 1992). High concentrations of *bcl-2* will also protect cells from apoptosis induced by *c-myc.*

Recently there has been much interest in the role of the tumor suppressor gene *p53* in apoptosis (Yonish-Rouach et al., 1991) and cancer (Chang et al., 1993). It has also been shown that in thymocytes there is a *p53*-dependent and independent pathway (Lowe et al., 1993; Clarke et al., 1993). One pathway initiated by DNA damage has an absolute requirement for the product of the *p53* gene. The other pathway initiated by glucocorticoids, aging, or calcium ionophores is a *p53*-independent pathway. Inactivation of the *p53* pathway will lead to the survival of cells with DNA damage, i.e., cells which are initiated. This may well result in an increased tumor incidence.

Apoptosis has also been shown to exert a major influence on tumor promotion, in particular in the liver. A number of chemicals including phenobarbitone, hexachlorocyclohexane, cyproterone acetate, and other steroid and peroxisomal proliferators induce liver hyperplasia. When treatment with the chemicals is stopped, regression of the hyperplasia occurs through apoptosis (Bursch et al., 1986). Cell death could be prevented by retreatment with the mitogens after withdrawal (Bursch et al., 1990). Tumor promoters, such as phenobarbitone, inhibit apoptosis in preneoplastic foci leading to growth of the foci and accelerated tumor development (Bursch et al., 1984, 1992).

Many cancer chemotherapeutic agents appear to kill tumor cells by apoptosis, so there is a great deal of interest in understanding the control of this process in both normal and tumor cells (Dive and Hickman, 1991; Sen and D'Incalci, 1992). It is hoped that such an understanding will lead to the development of new targets and strategies in cancer chemotherapy.

B. APOPTOSIS AND AIDS

AIDS is a complex disease induced by human immunodeficiency virus (HIV) infection. It has been proposed that many of the immunological and nonimmunological defects leading to AIDS may be related to an activation-induced apoptosis in CD4+ T cells and neurons (Ameisen, 1992). AIDS could be considered a pathological imbalance between the rate of CD4 cell death and replacement. Activation of CD4+ T cells from HIV-infected asymptomatic individuals, by calcium ionophore or antibodies to the T cell receptor, result in cell death by apoptosis (Groux et al., 1992). It has been suggested that the external glycoprotein gp 120 of the virus may bind to the CD4 receptors resulting in apoptosis of the normal CD4+ T cells on subsequent stimulation of the T cell receptor (Banda et al., 1992). Other mechanisms including defects in activation signaling could also contribute to the induction of apoptosis in HIV infection (Gougeon and Montagnier, 1993). To date there has been very little progress in the treatment of AIDS patients. New avenues are currently being explored in treating this disease with combinations of different drugs including antivirals, antibiotics, and antiapoptotic agents (Gougeon and Montagnier, 1993). Currently there are many exciting advances in the field of apoptosis, and it is to be hoped that in the next few years such results will lead to either improved treatments or prevention of existing diseases.

REFERENCES

Ameisen, J. C., Programmed cell death and AIDS: from hypothesis to experiment, *Immunology Today,* 13, 388, 1992.

Arends, M. J., Morris, R. G., and Wyllie, A. H., Apoptosis: the role of the endonuclease, *Am. J. Pathol.,* 136, 593, 1990.

Arends, M. J. and Wyllie, A. H., Apoptosis: mechanisms and roles in pathology, *Int. Rev. Exp. Path.,* 32, 223, 1991.

Aw, T. Y., Nicotera, P., Manzo, L., and Orrenius, S., Tributyltin stimulates apoptosis in rat thymocytes, *Arch. Biochem. Biophys.,* 283, 46, 1990.

Banda, N. K., Bernier, J., and Kurahara, D. K., et al., Crosslinking CD4 by human immunodeficiency virus gp 120 primes T cells for activation induced apoptosis, *J. Exp. Med.,* 38, 252, 1992.

Barres, B. A., Hart, I. K., Coles, H. S. R., Burne, J. F., Voyvodic, J. T., Richardson, W. D., and Raff, M. C., Cell death and control of cell survival in the oligodendrocyte lineage, *Cell,* 70, 31, 1992.

Barry, M. A. and Eastman, A., Endonuclease activation during apoptosis: the role of cytosolic Ca^{2+} and pH, *Biochem. Biophys. Res. Commun.,* 186, 782, 1992.

Brown, D. G., Sun, X.-M., and Cohen, G. M., Dexamethasone-induced apoptosis involves cleavage of DNA to large fragments prior to internucleosomal fragmentation, *J. Biol. Chem.,* 268, 3037, 1993.

Bursch, W., Düsterberg, B., and Schulte-Hermann, R., Growth, regression and cell death in rat liver as related to tissue levels of the hepatomitogen cyproterone acetate, *Arch. Toxicol.,* 59, 221, 1986.

Bursch, W., Lauer, B., Timmermann-Trosiener, I., Barthel, G., Schuppler, J., and Schulte-Hermann, R., Controlled death (apoptosis) of normal and putative preneoplastic cells in rat liver following withdrawal of tumor promoters, *Carcinogenesis,* 5, 453, 1984.

Bursch, W., Oberhammer, F., and Schulte-Hermann, R., Cell death by apoptosis and its protective role against disease, *TIPS,* 13, 245, 1992.

Bursch, W., Paffe, S., Putz, B., Barthel, G., and Schulte-Hermann, R., Determination of the length of the histological stages of apoptosis in normal liver and in altered hepatic foci of rats, *Carcinogenesis,* 11, 847, 1990.

Carson, D. A. and Ribeiro, J. M., Apoptosis and disease, *Lancet,* 341, 1251, 1993.

Chang, F., Syrjänen, S., Tervahauta, A., and Syrjänen, K., Tumourigenesis associated with the p53 tumour suppressor gene, *Br. J. Cancer,* 68, 653, 1993.

Clarke, A. R., Purdie, C. A., Harrison, D. J., Morris, R. G., Bird, C. C., Hooper, M. L., and Wyllie, A. H., Thymocyte apoptosis induced by p53-dependent and independent pathways, *Nature,* 362, 849, 1993.

Cohen, G. M., Sun, X.-M., Snowden, R. T., Dinsdale, D., and Skilleter, D. N., Key morphological features of apoptosis may occur in the absence of internucleosomal DNA fragmentation, *Biochem. J.,* 286, 331, 1992b.

Cohen, G. M., Sun. X.-M., Snowden, R. T., Ormerod, M. G., and Dinsdale, D., Identification of a transitional preapoptotic population of thymocytes, *J. Immunol.,* 151, 566, 1993.

Cohen, J. J. and Duke, R. C., Glucocorticoid activation of a calcium-dependent endonuclease in thymocyte nuclei leads to cell death, *J. Immunol.,* 132, 38, 1984.

Cohen, J. J., Duke, R. C., Fadok, V. A., and Sellins, K. S., Apoptosis and programmed cell death in immunity, *Ann. Rev. Immunol.,* 10, 267, 1992a.

Dive, C. and Hickman, J. A., Drug-target interactions: only the first step in the commitment to a programmed cell death?, *Br. J. Cancer,* 64, 192, 1991.

Ellis, R. E., Yuan, J., and Horvitz, H. R., Mechanisms and functions of cell death, *Ann. Rev. Cell Biol.,* 7, 663, 1991.

Evan, G. I. and Littlewood, T. D., The role of *c-myc* in cell growth, *Current Opinion in Genetics and Development,* 3, 44, 1993.

Evan, G., Wyllie, A., Gilbert, C., Littlewood, T., Land, H., Brooks, M., Waters, C., Penn, L., and Hancock, D., Induction of apoptosis in fibroblasts by *c-myc* protein, *Cell,* 63, 119, 1992.

Filipski, J., Leblanc, J., Youdale, T., Sikorska, M., and Walker, P. R., Periodicity of DNA folding in higher order chromatin structures, *EMBO J.,* 9, 1319, 1990.

Gaido, M. L. and Cidlowski, J. A., Identification, purification, and characterisation of a calcium-dependent endonuclease (NUC18) from apoptotic rat thymocytes, *J. Biol. Chem.,* 266, 18580, 1991.

Gerschenson, L. E. and Rotello, R. J., Apoptosis: a different type of cell death, *FASEB J.,* 6, 2450, 1992.

Gougeon, M.-L. and Montagnier, L., Apoptosis in AIDS, *Science,* 260, 1269, 1993.

Groux, H., Torpier, G., Monté, D., Mouton, Y., Capron, A., and Ameisen, J. C., Activation-induced death by apoptosis in CD4+ T cells from human immunodeficiency virus-infected asymptomatic individuals, *J. Exp. Med.,* 175, 331, 1992.

Hockenbery, D. M., Nuñez, G., Milliman, C., Schreiber, R. D., and Korsmeyer, S. J., Bcl-2 is an inner mitochondrial membrane protein that blocks programmed cell death, *Nature,* 348, 334, 1990.

Hockenbery, D. M., Oltvai, Z. N., Yin, Z.-M., Milliman, C. L., and Korsmeyer, S. J., Bcl-2 functions in an antioxidant pathway to prevent apoptosis, *Cell,* 75, 241, 1993.

Kerr, J. F. R., Shrinkage necrosis: a distinct mode of cellular death, *J. Pathol.,* 105, 13, 1971.

Kerr, J. F. R., Searle, J., Harmon, B. V., and Bishop, C. J., Apoptosis, in *Perspectives on Mammalian Cell Death,* Potten, C. S., Ed., Oxford University Press, New York, 1987, 93.

Kerr, J. F. R., Wyllie, A. H., and Currie, A. R., Apoptosis: a basic biological phenomenon with wide-ranging implications in tissue kinetics, *Br. J. Cancer,* 26, 239, 1972.

Korsmeyer, S. J., Bcl-2 initiates a new category of oncogenes: regulators of cell death, *Blood,* 80, 879, 1992.

Lockshin, R. A. and Zakeri, Z., Programmed cell death and apoptosis, in *Apoptosis: The Molecular Basis of Cell Death,* Tomei, L. D. and Cope, F.O., Eds., Cold Spring Harbor Laboratory, New York, 1991, 47.

Lowe, S. W., Schmitt, E. M., Smith, S. W., Osborne, B. A., and Jacks, T., p53 is required for radiation-induced apoptosis in mouse thymocytes, *Nature,* 362, 847, 1993.

Martin, S. J., Apoptosis: suicide execution or murder? *Trends in Cell Biology,* 3, 141, 1993.

Oberhammer, F., Fritsch, G., Pavelka, M., Froschl, G., Tiefenbacher, R., Purchio, T., and Schulte-Hermann, R., Induction of apoptosis in cultured hepatocytes and in the regressing liver by transforming growth factor-β1 occurs without activation of an endonuclease, *Toxicology Letters,* 64/65, 701, 1992.

Oren, M., The involvement of oncogenes and tumor suppressor genes in the control of apoptosis, *Cancer and Metastasis Reviews,* 11, 141, 1992.

Peitsch, M. C., Polzar, B., Stephan, H., Crompton, T., Robson MacDonald, H., Mannherz, H. G., and Tschopp, J., Characterization of the endogenous deoxyribonuclease involved in nuclear DNA degradation during apoptosis (programmed cell death), *EMBO J.,* 12, 371, 1993.

Raffray, M. and Cohen, G. M., Bis(tri-n-butyltin)oxide induces programmed cell death (apoptosis) in immature rat thymocytes, *Arch. Toxicol.,* 65, 135, 1991.

Sen, S. and D'Incalci, M., Apoptosis: biochemical events and relevance to cancer chemotherapy, *FEBS Lett.,* 307, 122, 1992.

Sentman, C. L., Shutter, J. R., Hockenbery, D., Kanagawa, O., and Korsmeyer, S. J., Bcl-2 inhibits multiple forms of apoptosis but not negative selection in thymocytes, *Cell,* 67, 879, 1991.

Smith, C. A., Williams, G. T., Kingston, R., Jenkinson, E. J., and Owen, J. J. T., Antibodies to CD3/T-cell receptor complex induce death by apoptosis in immature T cells in thymic cultures, *Nature,* 337, 181, 1989.

Spencer, C. A. and Groudine, M., Control of c-*myc* regulation in normal and neoplastic cells, *Adv. Cancer Res.,* 56, 1, 1991.

Strasser, A., Harris, A. W., and Cory, S., bcl-2 transgene inhibits T cell death and perturbs thymic self-censorship, *Cell,* 67, 889, 1991.

Sun, X.-M., Snowden, R. T., Dinsdale, D., Ormerod, M. G., and Cohen, G. M., Changes in nuclear chromatin precede internucleosomal DNA cleavage in the induction of apoptosis by etoposide, *Biochem. Pharmacol.,* 47, 187, 1994.

Sun, X.-M., Snowden, R. T., Skilleter, D. N., Dinsdale, D., Ormerod, M. G., and Cohen, G. M., A flow cytometric method for the separation and quantitation of normal and apoptotic thymocytes, *Anal. Biochem.,* 204, 351, 1992.

Thornberry, N. A. et al., A novel heterodimeric cysteine protease is required for interleukin-1β processing in monocytes, *Nature,* 356, 768, 1992.

Tomei, L. D., Shapiro, J. P., and Cope, F. O., Apoptosis in C3H/10T$^1/_2$ mouse embryonic cells: evidence for internucleosomal DNA modification in the absence of double-strand cleavage, *Proc. Natl. Acad. Sci. U.S.A.,* 90, 853, 1993.

Ucker, D. S., Obermiller, P. S., Eckhart, W., Apgar, J. P., Berger, N. A., and Meyers, J., Genome digestion is a dispensable consequence of physiological cell death mediated by cytotoxic T lymphocytes, *Mol. Cell Biol.,* 12, 3060, 1992.

Ushikubi, F., Aiba, Y.-I., Nakamura, K.-I., Namba, T., Hirata, M., Mazda, O., Katsura, Y., and Narumiya, S., Thromboxane A2 receptor is highly expressed in mouse immature thymocytes and mediates DNA fragmentation and apoptosis, *J. Exp. Med.,* 178, 1825, 1993.

Vaux, D. L., Cory, S., and Adams, J. M., Bcl-2 promotes the survival of haemopoietic cells and cooperates with c-myc to immortalize pre-B cells, *Nature,* 335, 440, 1988.

Vaux, D. L., Toward an understanding of the molecular mechanisms of physiological cell death, *Proc. Natl. Acad. Sci. U.S.A.,* 90, 786, 1993.

Walker, N. I., Harmon, B. V., Gobé, G. C., and Kerr, J. F. R., Patterns of cell death, *Meth. Achiev. Exp. Pathol.,* 13, 18, 1988.

Walker, P. R., Smith, C., Youdale, T., Leblanc, J., Whitfield, J. F., and Sikorska, M., Topoisomerase II-reactive chemotherapeutic drugs induce apoptosis in thymocytes, *Cancer Res.,* 51, 1078, 1991.

Waring, P., Egan, M., Braithwaite, A., Müllbacher, A., and Sjaarda, A., Apoptosis induced in macrophages and T blasts by the mycotoxin sporidesmin and protection by Zn^{2+} salts, *Int. J. Immunopharmac.,* 12, 445, 1990.

Waring, P., Kos, F. J., and Müllbacher, A., Apoptosis or programmed cell death, *Medicinal Res. Rev.,* 11, 219, 1991.

Williams, G. T. and Smith, C. A., Molecular regulation of apoptosis: genetic controls on cell death, *Cell,* 74, 777, 1993.

Wyllie, A. H., Glucocorticoid-induced thymocyte apoptosis is associated with endogenous endonuclease activation, *Nature,* 284, 555, 1980.

Wyllie, A. H., Kerr, J. F. R., and Currie, A. R., Cell death: the significance of apoptosis, *Int. Rev. Cytol.,* 68, 251, 1980.

Wyllie, A. H. and Morris, R. G., Hormone-induced cell death: purification and properties of thymocytes undergoing apoptosis after glucocorticoid treatment, *Am. J. Pathol.,* 109, 78, 1982.

Wyllie, A. H., Morris, R. G., Smith, A. L., and Dunlop, D., Chromatin cleavage in apoptosis: association with condensed chromatin morphology and dependence on macromolecular synthesis, *J. Pathol.,* 142, 67, 1984.

Yonish-Rouach, E., Resnitzky, D., Lotem, J., Sachs, L., Kimchi, A., and Oren, M., Wild-type p53 induces apoptosis of myeloid leukaemic cells that is inhibited by interleukin 6, *Nature,* 352, 345, 1991.

Yuan, J., Shanham, S., Ledoux, S., Ellis, H. M., and Horvitz, H. R., The C. elegans cell death gene *ced*-3 encodes a protein similar to mammalian interleukin-1β-converting enzyme, *Cell,* 75, 641, 1993.

Chapter 11

APOPTOSIS: SOME ASPECTS OF REGULATION AND ROLE IN CARCINOGENESIS

Rolf Schulte-Hermann

CONTENTS

I. Introduction 207

II. Types and Occurrence of Active Cell Death 208

III. Regulation of Active Cell Death 211

IV. Toxicological Implications 213

V. Summary 215

References 215

I. INTRODUCTION

Cell death has traditionally been considered by most pathologists and toxicologists as a passive degenerative phenomenon consequent to toxic injury. Some decades ago scientists began to recognize the occurrence and relevance of a type of cell death that could be actively induced by the organism (Farber et al., 1972; Glücksmann, 1951; Kerr, 1971; Lanzerotti and Gullino, 1972; Lockshin and Williams, 1965; Lockshin and Beaulaton, 1974; Saunders, 1966; Schweichel and Merker, 1973). In 1972 Kerr, Wyllie, and Currie introduced the new term "apoptosis"* and proposed a classification of cell death into two broad categories (Kerr et al., 1972). Apoptosis was understood as an active, inherently programmed phenomenon (Table 11.1). On the other hand, the old term "necrosis" was redefined to indicate cell death after massive tissue injury leading to rapid incapacitation of major cell functions. Other groups independently became interested in the study of active cell death as occurring during regression of hormone-dependent tumors after withdrawal of estrogens or during embryogenesis (Glücksmann, 1951; Lanzerotti and Gullino, 1972; Saunders, 1966; Schweichel and Merker, 1973). In nonvertebrates a nontoxic

---

* The term apoptosis is taken from Greek, where it was and is still used to indicate "falling down," e.g., of leaves from trees, or hair from the head. Necrosis means (cell) death.

type of cell death has long been known to occur during metamorphosis, where it functions in morphogenesis and is governed by the developmental program ("programmed cell death") (Lockshin and Williams, 1965; Lockshin and Beaulaton, 1974). Nevertheless, the concept of active cell death has been accepted reluctantly by the scientific community, and it is only recently that an upsurge of papers dealing with apoptosis has appeared.

Frequently, programmed cell death, apoptosis, etc., are used synonymously. This is not correct in our view; I will therefore use the term "active cell death" as proposed by Tenniswood (Tenniswood et al., 1992) (see below) (Table 11.1).

II. TYPES AND OCCURRENCE OF ACTIVE CELL DEATH

The definition of apoptosis has been based on morphological and functional grounds. Its characteristics are condensation of the cell and of nuclear chromatin at the nuclear membrane, fragmentation of the nucleus and the whole cell giving rise to apoptotic bodies (AB), and, finally, phagocytosis by phagocytes and neighboring epithelial cells. Cellular organelles, including the cell membrane, appear to be well preserved during apoptosis, and usually no inflammatory responses are seen around apoptotic cells. Lysosomes do not appear to play a role during early stages of apoptosis, but are involved in the degradation of the phagocytosed apoptotic bodies by the host cell (Kerr, 1971; Kerr et al., 1972).

TABLE 11.1
Characteristic Features of Active Cell
Death and Necrosis

Active Cell Death (a.c.d.)
(natural, physiological c.d.)

A process of active self-destruction of a cell, genetically encoded
Complementary to but opposite of mitosis
Serves to eliminate "unwanted" (damaged, precancerous, excessive) cells
Serves in embryogenesis at defined stages of the developmental program
    (programmed cell death)
Characteristic morphology allows distinction of following types:
        Condensation c.d. = apoptosis
        Lysosomal or autophagic c.d.
        Nonlysosomal disintegration

Necrosis

Cell death after violent environmental perturbation which leads to rapid
    breakdown of major cell functions (gene expression, ATP synthesis,
    membrane potential)

TABLE 11.2
Active Cell Death: Occurrence under Physiological Conditions (examples)

Ontogenesis

Nematodes	Maturation
Insects	Metamorphosis
Mammals	Embryo implantation: elimination of decidual cells
	Regression of sexual changes (e.g., Mullerian duct)
	Brain development

Adult Organisms

Lymphocytes	Elimination of self-reactive clones, termination of immune response
Uterus	Regression post partum
Vagina	Epithelial cells (stratum spinosum, stratum basalis) during late metestrus and early diestrus
Prostate	Castration
Liver	Normal cell turnover
	Regression of mitogen-induced hyperplasia
	Elimination of preneoplastic cells
Tumors (hormone-dependent)	Withdrawal of growth supporting hormone

Apoptosis is not the only form of active cell death. Early studies revealed a steep increase of lysosomal enzymes in hormone-dependent mammary cancer after ovariectomy, which does not fit the concept of apoptosis (Lanzerotti and Gullino, 1972). At the same time, Schweichel and Merker (1973) recognized the existence of morphologically different modes of active cell death during embryonal development; their "type I" cell death is characterized by cell and chromatin condensation and is probably identical to apoptosis. Type II is characterized by lysosomal activation and prominent autophagy of dying cells (Table 11.1) (Schweichel and Merker, 1973). A similar appearance of cell death has also been observed during metamorphosis in certain insect organs (Lockshin and Williams, 1965; Lockshin and Beaulaton, 1974; Zakeri et al., 1993). More recently these different types of cell death have been found during development of the nervous system, and evidence suggesting even more types of cell death has been presented (Clarke, 1990).

Active cell death is a phylogenetically old phenomenon which occurs in many situations in the embryo and adult (Table 11.2). In the adult organism one of the major functions is the maintenance of adequate size of organs. In rodent liver a variety of lipophilic xenobiotic compounds have been found to induce an adaptive type of organ growth by cellular hypertrophy and hyperplasia. Compounds capable of doing this include phenobarbital, hexachlorocyclohexane, cyproterone acetate (CPA), and peroxisome proliferators such as nafenopin, etc. When treatment with these agents is stopped after occurrence of liver growth, organ enlargement was found to be largely reversible, and, at least in some models, the hyperplasia readily disappeared by apoptosis within a few days. The occurrence of cell death could be stopped at any time by retreatment with a variety of liver mitogens indicating that indeed active cell death was taking place (Bursch et al., 1984, 1990, 1992a, b; Schulte-Hermann, 1985; Schulte-Hermann et al., 1990). Inhibition of active cell death by tissue-specific mitogens is an important characteristic and can be used to discriminate active cell death from passive phenomena (Table 11.3) (Bursch et al., 1992b). Other examples of tissue regression by active cell death or apoptosis in the adult organism are those provided by rat liver after lead nitrate (Columbano et al., 1985), the uterus post partum, the breast after weaning, adrenal cortex or thyroid after hypophysectomy, and prostate after castration.

The inhibition of apoptosis by mitogens can be used to estimate its duration. For this purpose we determined the kinetics of disappearance of the histological signs of apoptosis after mitogen treatment, using regression of CPA-induced liver hyperplasia as a model. A half-life of about 2 hours and an

TABLE 11.3
Active Cell Death: Involvement in Disease
(Examples) (see Bursch, 1992b for references)

Drug	Effect on cell death (inhibition/stimulation)	Disease
Teratogenesis		
Glucocorticoids	–	Cleft palate
Retinoids	+	Limb malformation
DES	–	Persistence of Mullerian duct, abnormalities and tumors
Immune Disease		
Glucocorticoids	+	Immunosuppression
2,3,7,8-TCDD	+	Immunosuppression
Tumorigenesis		
Phenobarbital	–	Liver tumor promotion (rodent)
Estrogens	–	Mammary tumors
bcl-2 gene over-expression	–	Burkitt's lymphoma
Other		
UV radiation	+	Sunburn
Hepatitis B virus	+	Hepatitis
Cytarabine	+	Acute liver injury
Irradiation	+	Acute tissue damage

average duration of 3 hours of the histologically visible part of apoptosis was calculated (Bursch et al., 1990). Such estimates allow computation of the rate of apoptosis from histological counts. In regressing liver and in preneoplastic liver foci (see below), rates of apoptosis (apoptotic bodies/100 hepatocytes/ hour) were 0–5% and 2–5%/hour, respectively. Such estimates should be useful for mathematically modeling growth kinetics of normal and neoplastic or preneoplastic tissues (Moolgavkar, 1989).

As mentioned above, early studies suggested the occurrence of type II cell death (lysosomal-autophagic) during regression of mammary tumors. We have studied the human mammary cancer cell line MCF 7. The growth of this cell line is strictly estrogen-dependent, and anti-estrogens such as tamoxifen, toremifen or ICI 164 384 inhibit DNA-synthesis and induce cell death (Bardon et al., 1987; Wärri et al., 1993; Bursch et al., submitted). Morphological investigations have shown little evidence of apoptotic morphology. Rather, dying cells exhibit an increase of lysosomes and degradation of cytoplasm suggesting involvement of type II cell death. Estrogen treatment blocks cell death induced by tamoxifen indicating an active nature of cell death (Bursch, Ellinger, Kienzl, Török, unpublished). A lysosomal type of cell death has also been seen in a number of other models (Clarke, 1990; Lanzerotti and Gullino, 1972; Lockshin and Beaulaton, 1974; Schwartz et al., 1993; Schweichel and Merker, 1973; Zakeri et al., 1993); so far it is not known what determines the type of cell death occurring in a tissue during involution.

As to the biochemistry of apoptosis, it has been suggested that the characteristic condensation of nuclear chromatin results from activation of an endonuclease and degradation of chromatin into mono- and oligonucleosomal fragments. Upon gel electrophoresis such fragmented chromatin yields a "DNA-ladder," which has become popular as a biochemical hallmark of apoptosis. However, DNA laddering was not found during early stages of apoptosis in hepatocytes or some other cell types (Oberhammer et al., 1993a, b, c; Zakeri et al., 1993). Recent evidence suggests that chromatin condensation during apoptosis is associated with formation of large chromatin fragments 50 and 300 kbp long (Filipski et al., 1990; Oberhammer et al., 1993b) and that nucleases different from those forming oligonucleosomal patterns are involved in this process (Brown et al., 1993). Further information on this point is presented in Chapter 10 by G. M. Cohen et al. in this volume.

III. REGULATION OF ACTIVE CELL DEATH

The concept of active cell death being a genetically encoded process has stimulated an intense search for the genes involved. So far several genes have been found to be associated with active cell death or apoptosis. Examples include *myc* or *fos,* which are also involved in cell replication and other states of increased functional performance of cells (Buttyan et al., 1988; Fanidi et al., 1992). Of great interest are some genes which apparently serve to inhibit cell death such as the *bcl-2* gene, overexpression of which prevents apoptosis of B-lymphocytes and may be involved in lymphoma formation (Hockenberry et al., 1990). A functional homologue, *ced 9,* inhibits programmed cell death in the development of the worm *Caenorhabditis elegans.* The oncogene *fas* codes for the TNF-receptor; activation of this receptor by TNF or by anti-FAS-antibodies (including antibodies to the APO-1 antigen) result in rapid apoptosis of the target cells (and potentially in the death of the organism) (Debatin et al., 1990; Ogasawara et al., 1993).

One of the first genes associated with apoptosis was the *TRPM 2* (or *clusterin*) gene, which is strongly expressed in prostate after castration and can be repressed by testosterone (TRPM = testosterone repressed prostate message) (Léger et al., 1988). *Clusterin/TRPM 2* was also found to be expressed in the liver during regression of mitogen-induced hyperplasia, but not in another state of massive cell elimination, namely after liver damage by CCL_4. However, when the *TRPM 2* expression was analyzed at the single cell level by *in situ* hybridization on histological liver sections, the label was more or less equally distributed among all hepatocytes. This observation made a specific role of *TRPM 2* in hepatocyte apoptosis unlikely (Bursch et al., 1995). Indeed, in recent years it became evident that *TRPM 2* exerts a variety of different functions in the organism. Hypothetically, its role in tissue regression might be to protect cells from complement-induced lysis. Prevention of lysis is considered characteristic of apoptosis. Likewise one may predict that dying cells may use many genes that also serve other functions (see involvement of *fos, myc,* etc., mentioned above). It remains to be learned whether specific apoptosis genes or "thanatogenes" do exist at all.

It would be of particular interest to know by what intercellular signals and factors apoptosis can be induced and controlled in a given cell. Obviously the means by which this control is effected must be part of the growth regulatory network which maintains homeostasis of cell number in accordance with the demands of the organism. Trophic hormones and growth factors will inhibit apoptosis (e.g., ACTH in adrenal cortex, testosterone in the prostate), while other factors will induce or enhance apoptotic activity, such as TNF (tumor necrosis factor), glucocorticoids in lymphatic cells, or thyroxin in tadpole tail during metamorphosis.

We became interested in transforming growth factor β1 (TGF β1), a member of a family of peptides with growth inhibitory functions, on many cells (Roberts et al., 1988). TGF β1 has been found to inhibit hepatocyte DNA synthesis *in vitro* and in regenerating liver *in vivo* (Russell et al., 1988). In our studies on regressing rat liver, apoptotic hepatocytes showed positive immunostaining for TGF β1 precursor (antibodies against the 266–278 portion of pre-TGF β1 were kindly provided by M. Sporn, Bethesda). Antibodies against the mature TGF β1 also gave a positive signal, but much weaker; this may be explained by the short half-life of the mature form (Bursch et al., 1993). TGF β1 also induced apoptosis *in vitro* in cultured hepatocytes and *in vivo* when injected as a pulse some hours before sacrifice, the dose being identical to that inhibiting DNA synthesis in regenerating liver (Oberhammer et al., 1992; Oberhammer et al., 1993a). Interestingly, TGF β1 was particularly effective in regressing liver after cessation of treatment with the liver mitogen CPA. Apparently it acts preferentially upon cells already committed to apoptosis by pre-existing hyperplasia. In other words, under these experimental conditions, TGF β1 would not appear to be a primary signal for apoptosis but rather a permissive factor. Another surprising finding was that also some seemingly intact hepatocytes stained positive for pre-TGF β1. Since the incidence of these positive cells always correlated with that of apoptosis we concluded that hepatocytes preparing for apoptosis start to show pre-TGF β1 2 to 3 hours before chromatin condensation (Bursch et al., 1993). This observation may provide a marker to identify early apoptotic cells. Furthermore, staining for pre-TGF β1 may be useful in discriminating apoptosis from necrosis because necrotic hepatocytes occurring after administration of liver toxins did not stain.

IV. TOXICOLOGICAL IMPLICATIONS

Given the importance of active cell death during development and in adult life, it is not surprising that chemicals can exert toxicity by interfering with the regulation of apoptosis. Examples are given in Table 11.3. As shown, chemicals can either stimulate cell death, thereby leading to a loss of important cells, or they may inhibit cell death where it physiologically should occur. Thus the concept of active cell death has provided new insights into the pathophysiology of various diseases. As an example, some findings related to carcinogenesis will be described below.

It is now generally accepted that carcinogenesis occurs through stages including initiation, promotion, progression, and finally tumor growth and metastasis (Pitot and Sirica, 1980; Schulte-Hermann, 1985). Cell replication plays an essential role in these stages; apoptosis seems to be similarly important. Multistage carcinogenesis can be studied particularly well in (rodent) liver where putative initiated cells can be detected at the single-cell stage and as small clones and large advanced lesions by histochemical and immunocytochemical techniques. These early lesions show much higher rates of DNA synthesis and mitosis and grow more rapidly than normal liver. Tumor promoters such as phenobarbital and hexachlorocyclohexane enhance the growth rate of these early lesions, but surprisingly, have little effect on DNA synthesis. The explanation appears to be that putative preneoplastic foci usually exhibit not only enhanced DNA synthesis but also enhanced apoptotic activity and that tumor promoters favor foci growth by inhibiting apoptosis. Obviously, the tumor promoters studied act like trophic hormones and mitogens in their target tissue (see above). Withdrawal of the model promoter phenobarbital resulted in a dramatic increase of apoptosis in foci. This also explains why promotion can be reversible (Bursch et al., 1984; Schulte-Hermann, 1985; Schulte-Hermann et al., 1990).

TABLE 11.4
Implications of Enhanced Level of Birth and Death of
Cells in Various Stages of Cancer Development

Initiation

Efficiency and persistence limited by death of initiated cells
Can initiation be reversible?

Promotion

Inhibition of cell death leads to accumulation of preneoplastic cells.
Withdrawal of promoter causes excessive cell death, thus explaining reversibilty of promotion.

Tumor growth

Mitogen withdrawal or administration of anti-hormones results in excessive cell death leading
to regression and potential elimination of tumor.

We have studied the number of putative initiated single cells directly by using glutathione S transferase P (GST-P) as a marker. Following a single toxic dose of N-nitrosomorpholine we observed a steep rise of single GST-P positive cells and (with some delay) of mini-foci in the liver. After a plateau phase the number of these single cells dramatically declined by approximately 80%, and there was evidence of apoptosis of these single cells (Grasl-Kraupp, Wagner, Ruttkay-Nedecky, Schulte-Hermann, manuscript in preparation). Interestingly, it has been previously predicted from mathematical models of carcinogenesis that initiated cells can be extinguished (Moolgavkar, 1989); this prediction fits well with the experimental results described above. These findings, if confirmed, will raise new insights and questions on carcinogenesis. One is that the effect of initiation in a tissue can be reversed by apoptosis (Grasl-Kraupp et al., 1994). Since apoptosis in putative preneoplastic foci was always much more frequent than in normal liver cells, it may serve as a defense line or guard for the organism against the development of cancer. Tumor promoters may undermine this defense by inhibiting apoptosis.

Finally, the growth rate of tumors is also, in large part, determined by the rate of apoptosis. Using an estrogen-dependent hamster kidney tumor as a model, we have previously shown that in this tumor both mitosis and apoptosis are under estrogen control. Withdrawal of estrogens resulted in a massive increase of apoptosis and a decline of mitosis, while the reverse was found upon retreatment with estrogen (Bursch et al., 1991). Likewise, in rat liver, benign and even malignant tumors induced by the peroxisome proliferator nafenopin largely disappeared upon cessation of treatment. The mechanism of regression again was a dual one: there was an increase in the incidence of cell death and a complete disappearance of DNA synthesis in the tumors (Grasl-Kraupp, Wagner, Ruttkay-Nedecky, Schulte-Hermann, manuscript in preparation).

An overview on the proposed roles of active cell death during the various stages of carcinogenesis is listed in Table 11.4. Hopefully, in the future these findings may prove useful in providing new preventive and therapeutic strategies in the fight against cancer.

V. SUMMARY

Apoptosis is a type of active cell death. Its discovery and major characteristics are briefly discussed. Genes and regulatory factors involved in the control of apoptosis are still incompletely known. One of the factors triggering apoptosis appears to be transforming growth factor $\beta 1$. Toxic chemicals can interfere with the regulation of active cell death, a discovery which provides new insight into the mechanism of various diseases. As an example, the role of apoptosis in (chemical) carcinogenesis is described in some detail.

REFERENCES

Bardon, S., Vignon, F., Montcourrier, P., and Rochefort, H., Steroid receptor-mediated cytotoxicity of an antiestrogen and an antiprogestin in breast cancer cells, *Cancer Res.,* 47, 1441, 1987.

Brown, D. G., Sun, X.-M., and Cohen, G. M., Dexamethasone-induced apoptosis involves cleavage of DNA to large fragments prior to internucleosomal fragmentation, *J. Biol. Chem.,* 268, 3037, 1993.

Bursch, W., Fesus, L., and Schulte-Hermann, R., Apoptosis ("programmed" cell death) and its relevance in liver injury and carcinogenesis, in *Tissue Specific Toxicology: Biochemical Mechanisms,* Dekant, W. and Neumann, H. G., Eds., Academic Press Ltd., London, 1992a, chap. 5.

Bursch, W., Gleeson, T., Kleine, L., and Tenniswood, M., Expression of clusterin (testosterone-repressed prostate message-2) mRNA during growth and regression of rat liver, *Arch. Toxicol.,* 69, 253, 1995.

Bursch, W., Kienzl, H., Ellinger A., and Schulte-Hermann, R., Cell death in cultured human mammary carcinoma cells (MCF-7) after treatment with the antiestrogens tamoxifen and ICI 164 384, submitted.

Bursch, W., Lauer, B., Timmermann-Trosiener, T., Barthel, G., Schuppler, J., and Schulte-Hermann, R., Controlled cell death (apoptosis) of normal and putative neoplastic cells in rat liver following withdrawal of tumor promoters, *Carcinogenesis,* 5, 53, 1984.

Bursch, W., Liehr, J. G., Sirbasku, D., Putz, B., Taper, H., and Schulte-Hermann, R., Control of cell death (apoptosis) by diethylstilbestrol in an estrogen dependent kidney tumor, *Carcinogenesis,* 12, 855, 1991.

Bursch, W., Oberhammer, F., Jirtle, R. L., Askari, M., Sedivy, R., Grasl-Kraupp, B., and Purchio, A. F., Transforming growth factor-$\beta 1$ as a signal for induction of cell death by apoptosis, *Br. J. Cancer,* 67, 531, 1993.

Bursch, W., Oberhammer, F., and Schulte-Hermann, R., Cell death by apoptosis and its protective role against disease, *Trends Pharmacol. Sci.,* 13, 245, 1992b.

Bursch, W., Paffe, S., Putz, B., Barthel, G., and Schulte-Hermann, R., Determination of the length of the histological stages of apoptosis in normal liver and in altered hepatic foci of rats, *Carcinogenesis,* 11, 847, 1990.

Buttyan, R., Zakeri, Z., Lockshin, R. A., and Wolgemuth, D., Cascade induction of c-fos, c-myc, and heat shock 70 k transcripts during regression of the rat ventral prostate gland, *Mol. Endocrinol.,* 2, 650, 1988.

Clarke, P. G. H., Developmental cell death: morphological diversity and multiple mechanisms, *Anat. Embryol.,* 181, 195, 1990.

Columbano, A., Ledda-Columbano, G. M., Coni, P. P., Faa, G., Liguori, C., Santa Cruz, G., and Pani, P., Occurrence of cell death (apoptosis) during the involution of liver hyperplasia, *Lab. Invest.,* 52, 670, 1985.

Debatin, K., Goldmann, C. K., Bamford, R., Waldmann, T. A., and Krammer, P. H., Monoclonal-antibody-mediated apoptosis in adult T-cell leukemia, *Lancet,* 335, 497, 1990.

Fanidi, A., Harrington, E. A., and Evan, G. I., Cooperative interaction between c-myc and bcl-2 proto-oncogenes, *Nature,* 359, 554, 1992.

Farber, E., Verbin, R. S., and Lieberman, M., Cell suicide and cell death, in *A Symposium on Mechanisms of Toxicology,* Aldridge, N., Ed., Macmillan, New York, 1972, 163.

Filipski, J., Leblanc, J., Youdale, T., Sikorska, M., and Walker, P. R., Periodicity of DNA folding in higher order chromatin structures, *EMBO J.,* 9, 1319, 1990.

Glücksmann, A., Cell death in normal vertebrate ontogeny, *Biol. Rev. Cambridge Phil. Soc.,* 26, 59, 1951.

Grasl-Kraupp, B., Bursch, W., Ruttkay-Nedecky, B., Wagner, A., Lauer, B., and Schulte-Hermann, R., Food restriction eliminates preneoplastic cells through apoptosis and antagonizes carcinogenesis in rat liver, *Proc. Nat. Acad. Sci. U.S.A.,* 91, 9995, 1994.

Hockenberry, D., Nunez, G., Milliman, C., Schreiber, R. D., and Korsmeyer, S. J., Bcl-2 is an inner mitochondrial membrane protein that blocks programmed cell death, *Nature,* 348, 334, 1990.

Kerr, J. F. R., Shrinkage necrosis: a distinct mode of cellular death, *J. Pathol.,* 105, 13, 1971.

Kerr, J. F. R., Wyllie, A. H., and Currie, A. R., Apoptosis: a basic biological phenomenon with wide-ranging implications in tissue kinetics, *J. Cancer,* 26, 239, 1972.

Lanzerotti, L. H. and Gullino, P. M., Activity and quantity of lysosomal enzymes during mammary tumor regression, *Cancer Research,* 32, 2679, 1972.

Léger, J., Le Guellec, R., and Tenniswood, P. R., Treatment with antiandrogens induces an androgen repressed gene in the rat ventral prostate, *The Prostate,* 13, 131, 1988.

Lockshin, R. A. and Beaulaton, J., Programmed cell death. cytochemical evidence for lysosomes during the normal breakdown of the intersegmental muscles, *J. Ultrastruct. Res.,* 46, 43, 1974.

Lockshin, R. A. and Williams, C. M., Programmed cell death. I. Cytology of degeneration in the intersegmental muscles of the Pernyi silk moth, *J. Insect. Physiol.,* 11, 123, 1965.

Moolgavkar, S. H., Multistage models for cancer risk assessment, in *Biologically Based Models for Cancer Risk Assessment,* Travis, C. C., Ed., Plenum Press, New York, 1989, 9.

Oberhammer, F., Bursch, W., Tiefenbacher, R., Fröschl, G., Pavelka, M., Purchio, T., and Schulte-Hermann, R., Apoptosis is induced by transforming growth factor-β1 within 5 hours in regressing liver without significant fragmentation of the DNA, *Hepatology,* 18, 1238, 1993a.

Oberhammer, F., Fritsch, G., Schmied, M., Pavelka, M., Printz, D., Purchio, T., Lassmann, H., and Schulte-Hermann, R., Condensation of the chromatin at the membrane of an apoptotic nucleus is not associated with activation of an endonuclease, *J. Cell Sci.,* 104, 317, 1993c.

Oberhammer, F., Pavelka, M., Sharma, S., Tiefenbacher, R., Purchio, T. A., Bursch, W., and Schulte-Hermann, R., Induction of apoptosis in cultured hepatocytes and in regressing liver by transforming growth factor-β1, *Proc. Nat. Acad. Sci. U.S.A.,* 89, 5408, 1992.

Oberhammer, F., Wilson, J. W., Dive, C., Morris, I. D., Hickman, J. A., Wakeling, A. E., Walker, P. R., and Sikorska, M., Apoptotic death in epithelial cells: cleavage of DNA to 300 and/or 50 kb fragments prior to or in the absence of internucleosomal fragmentation, *EMBO J.,* 12, 3679, 1993b.

Ogasawara, J., Watanabe-Fukunaga, R., Adachi, M., Matsuzawa, A., Kasugai, T., Kitamura, Y., Itoh, N., Suda, T., and Nagata, S., Lethal effect of the anti-Fas antibody in mice, *Nature,* 364, 806, 1993.

Pitot, H. C. and Sirica, A. E., The stages of initiation and promotion in hepatocarcinogenesis, *Biophys. Acta,* 605, 191, 1980.

Roberts, A. B., Thompson, N. L., Heine, U., Flanders, C., and Sporn, M. B., Transforming growth factor-β: possible roles in carcinogenesis, *Br. J. Cancer,* 57, 594, 1988.

Russell, W. E., Coffex, R. J., Jr., Ouellette, A. J., and Moses, H. L., Type β transforming growth factor reversibly inhibits the early proliferative response to partial hepatectomy in the rat, *Proc. Nat. Acad. Sci. U.S.A.,* 85, 5126, 1988.

Saunders, J. W., Jr., Death in embryonic systems, *Science,* 154, 604, 1966.

Schulte-Hermann, R., Tumor promotion in the liver, *Arch. Toxikol.,* 57, 147, 1985.

Schulte-Hermann, R., Timmermann-Trosiener, I., Barthel, G., and Bursch, W., DNA synthesis, apoptosis and phenotypic expression as determinants of growth of altered foci in rat liver during phenobarbital promotion, *Cancer Research,* 50, 5127, 1990.

Schwartz, L. M., Smith, S. W., Jones, M. E. E., and Osborne, B. A., Do all programmed cell deaths occur via apoptosis? *Proc. Nat. Acad. Sci.,* 90, 980, 1993.

Schweichel, J.-U. and Merker, H. J., The morphology of various types of cell death in prenatal tissues, *Teratology,* 7, 253, 1973.

Tenniswood, M. P., Guenette, R. S., Lakins, J., Mooibroek, M., Wong, P., and Welsh, J.-E. Active cell death in hormone-dependent tissues, *Cancer Metatases Rev.,* 11, 197, 1992.

Wärri, A. M., Huovinen, R. L., Laine, A. M., Martikainen, P. M., and Härkönen, P. L., Apoptosis in toremifene-induced growth inhibition of human breast cancer cells in vivo and in vitro, *J. Nat. Cancer Inst.,* 17, 1412, 1993.

Zakeri, Z. F., Quaglino, D., Latham, T., and Lockshin, R. A., Delayed internucleosomal DNA fragmentation in programmed cell death, *FASEB J.,* 7, 470, 1993.

INDEX

A

Accelerator mass spectrometry, 109–110
Adduct formation, see DNA adducts; Protein adducts
Adriamycin, 40, 42
Aflatoxins
  metabolism, 70–74
    in humans, 73–74
    in rodents, 70–73
  monitoring human exposure, 76–77
  oncogenes and tumor suppressor genes, 77–78
  resistance mechanisms, 75–76
  species susceptibility, 75
AIDS, apoptosis and, 202–203
Albumin-carcinogen adducts, 112–113
Alkylating agents, 132
Alkyltin, 189, see also Triorganotin
Amino acids
  neurotransmitters, 25
  oxidative damage to proteins, 49
  protein-carcinogen adducts, 113–115
Amosite, 87, 88
Amphibole asbestos, 87
Amyotrophic lateral sclerosis, parkinsonism, and dementia complex, 25, 26
Animal models, see also specific toxins
  aflatoxin susceptibility, 70–73, 75
  genetic variation in, 165–180
    increasing, experimental design for, 176–180
    new genetics, relevance to toxicological research, 175–176
    response to xenobiotics, 166–167
    strain differences as research tool, 173–175
    toxicological screening systems, 171–175
    variability, humans versus animals, 168–171
Anoxia-reperfusion injury, 39, 56
Antioxidants, 37
Apoptosis, 185–203
  agents inducing, 189
  biochemical characteristics, 187–188
  in carcinogenesis, 207–215
    regulation of, 211–213
    toxicological implications, 213–215

types and occurrence of active cell death, 208–211
  and disease, 199–203, 207–215
  DNA degradation, 189–195
    dissociation of internucleosomal cleavage from, 190–193
    formation of large fragments before internucleosomal cleavage, 193–195
  genes involved in, 188–189
  in situ hybridization applications, 158–159
  morphologic changes, 186
  necrosis versus, 185–187
  occurrence of, 187
  thymocytes
    model of, 199
    preapoptotic populations, 195–199
Arachidonic acid, 48
Archived tissues, DNA preservation in, 148–150
Asbestos, 85–88, 94
Ascorbic acid, 37, 42–47, 54
AT125, 71
ATP
  mitochondrial permeability transition, 58, 60
  oxidative stress and, 49
  phosphorus-32 postlabeling, 105–108
ATP synthase, 8
Atractyloside, 58

B

Biochemistry, see Enzymes; Metabolism
Biomarkers, carcinogen, 103
1,3-Bis(2-chloroethyl)-1-nitrosourea (BCNU), 52, 54
Blood-brain barrier, 4–5
Bongokeric acid, 58
Brain, see Neurotoxicity
Brucite, 88
tert-Butyl hydroperoxide, 56–58

C

Calcium, 49
  and oxidative stress, 37, 56–60
  and hypoxic injury, 39

Carbonylcyanide-m-chlorophenylhydrozone, 54
Carcinogens
    aflatoxin, 69–78
    apoptosis and, 201–202, 207–215
    biomonitoring and molecular dosimetry, 101–117
        background levels, 101
        DNA adducts, 103–111
        dose-response relationship, 115–116
        protein adducts, 111–116
    DNA interactions, 125–138
        adduct formation in vivo, 133–138
        covalent binding to DNA, discovery of, 126–128
        genotoxic and nongenotoxic carcinogens, 132–133
        theories of chemical carcinogenesis, 128–132
    fibers, mineral, 85–95
Cell membranes, see Membranes
Cell biology
    apoptosis, see Apoptosis
    in situ hybridization applications, 156–158
    oxidative damage during redox cycling, 36, 40–42
Ceramic fibers, 90–93
Chelators, 54, 58
Chemotherapy, 137
Chlorinated phenols, 7
Chlorodinitrobenzene (CDNB), 76
Choreathetosis syndrome, 9, 10, 21, 22
Chrysotile, 87, 88, 91, 92
Cismethrin, 9, 22
CPP, 26
Crocidolite, 87
CS (choreathetosis) syndrome, 9, 10, 21, 22
Cyanopyrethroids, 21, 22
Cyclosporin A, 56–60
Cysteine, 49, 50, 113–116
Cytochrome P-450, 136
    aflatoxin susceptibility, 70
    neurotoxin metabolism, 5
Cytosolic factor, 57, 58

D

Dacarbazine, 137
Delayed neuropathy, organophosphorus-induced, 11–13, 26–30
Deltamethrin, 9, 22
Desferrioxamine, 54

Dexamethasone, 191–195, 198
Dichlorvos, 28
Diisopropylfluorophosphate (DFP), 28
N,N'-Diisopropyl phosphoramido-fluoridate (Mipafox), 9, 10
Dimethylnitrosamine, 127
Dinitrophenol, 7
N,N'-Diphenyl-p-phenylenediamine, 54
Diquat, 40, 41
Disproportionation of superoxide radical, 38
DNA
    hybridization, see *In situ* hybridization
    in apoptosis, 187–202
DNA adducts, 103–111, 115–116
    aflatoxin and, 76
    asbestos and, 94
    and carcinogenesis, 125–138
        adduct formation in vivo, 133–138
        covalent binding of carcinogens, discovery of, 126–128
        genotoxic and nongenotoxic carcinogens, 132–133
        theories of chemical carcinogenesis, 128–132
Dopamine depletion, 24

E

EDTA, 54
EGTA, 54
Electron transport, mitochondrial permeability transition, 60
ELISA, 108, 112
Enzyme-linked immunosorbent assays, 108, 112
Enzymes
    aflatoxin susceptibility, 70–74
    esterase inhibitor-induced neuropathy, 11–13, 26–30
    monoamine oxidases, 24
    neurotoxin metabolism, 5
    reactive oxygen species, 38
Epigenetic carcinogens, 133
EQ, 76
Erionite, 89
Etoposide, 189, 194
Excitatory amino acids, 25
Exposure markers, DNA adducts as, 135–137

F

Fatty acids, and mitochondrial permeability transition, 57

Fenton reaction, 38, 39
Fibers, mineral, 85–95
    asbestos, early studies, 85–88
    man-made, 89–93
    non-asbestos, 88–89
    pathogenetic mechanisms, 93–95
Fluorescence assays, 109
fos, 211
Free radicals, see Oxygen toxicity

G

Genetics, see also *In situ* hybridization
    aflatoxin susceptibility, 77–78
    apoptosis, 188–189
    regulation of apoptosis, 211–213
Genetic variation, see also Animal models
    laboratory animal populations, 168–171
    response to xenobiotics, 166–167
    strain differences as research tool,
        173–175
    types of, humans versus animals,
        173–174
Genotoxins, 132–133, see also Carcinogens;
        DNA adducts
GGA to GAA transition, 130
Glass fiber, 90, 92
Glucocorticoids, 189
Glutamate metabolism, 26
Glutathione metabolism, 37, see also
        Oxygen toxicity
    aflatoxin susceptibility, 70–76
    calcium as factor in oxidative stress,
        53–55
    mitochondrial permeability transition, 59,
        60
    nitrofurantoin-induced cytotoxicity, 49
    as pro-oxidant, 44
    and protein sulfhydryl groups, 39
    and α-tocopherol-ascorbic acid
        antioxidant activity, 44, 46, 47
Glutathione peroxidase, 39, 41
Glutathione peroxidase gene, 36
Glutathione-*S*-transferase, 213
Granulocytes, 40
Guam disease, 25

H

Haber-Weiss reaction, 38, 39
Ha-ras, 102, 130
Heavy metals, and mitochondrial
        permeability transition, 57

Hemoglobin-carcinogen adducts, 113–116
Hepatitis B virus, 77
2,5-Hexanedione, 30
Histone mRNA, 156–158
Humans
    aflatoxin metabolism in, 73–74
    genetic variation, 168, 173–174
Hydrogen peroxide, 37–40, 50–51
Hydroxyl radical scavengers, 42
1-Hydroxypyrene, 102
Hypoxia-reperfusion injury, 39, 56

I

Immunoassays
    DNA-carcinogen adducts, 108
    protein-carcinogen adducts, 111
Immunoglobulin light chain mRNA,
        155–156
In situ hybridization, 147–161
    applications, 156–161
    DNA preservation in archived tissues,
        148–149, 150
    immunoglobulin light chain mRNA as
        model system, 155–156
    technique, 150–155
Internucleosomal cleavage, 190–195
Ion channels, 21
Iron, 42, 43–44, 54
Ischemia-reperfusion injury, 39, 56

K

Kaolin RCF, 92
Kidney, neurotoxin metabolism, 5
Ki-ras, 77, 130
Koch's Postulates, 128
Korean tremolite, 88

L

Laboratory animals, see Animal models
Lead compounds, 5, 7, 8
Leukocytes, 40
Lipid peroxidation, 36
    antioxidant interactions against oxygen
        toxicity, 42–49
    glutathione redox cycle, 51–53, 56
Lipophilicity, neurotoxin, 5
Liposome peroxidation, 45
Liver
    aflatoxin susceptibility, 70–78
    neurotoxin metabolism, 5

M

Macrophages, 90
Malondialdehyde (MDA), 43, 51, 53
MAO-B, 24
Mass spectrometry, 109–110, 111, 114, 127
Mechloroethamine, 137
Membranes
   in apoptosis, 190
   mitochondrial permeability transition, 57–60
   oxidative damage, 42, 44, 47, 48, 56
Menadione, 42
Mephenesin, 22
Mesothelioma, 85–89
Messenger RNA, see *In situ* hybridization
Metabolism
   aflatoxin susceptibility, 70–75
   in apoptosis, 187–188, 194
   and carcinogenicity, 102
   genetic control and variability, 166–167
Metals, see also Transition metals
   and mitochondrial permeability transition,
      57, 58
   neurotoxicity, 5–8
Methamidophos, 28
4,4'-Methylene dianiline (MDA), 102
Methylguanine, 136, 138
N-Methylnitrosourea, 102, 131
Microsomal lipid peroxidation, 47
Mineral fibers, see Fibers, mineral
Mineral wool, 92
Mipafox (*N, N'*-diisopropyl
      phosphoramidofluoridate), 9, 10
Mitochondria
   dietary vitamin E and, 47–48
   glutathione, 55–56
   glutathione redox cycle, 56
   hypoxic myocardium, 39
   oxidative damage to membranes, 42
   permeability transition, 57–60
MK-801, 26
Monoamine oxidases, 24
MOPP, 137
MPP (1-methyl-4-phenylpyridinium), 24–26
MPTP (1-methyl-4-phenyl-1,2,3,6-
      tetrahydropyridine), 20–21, 25
Mutagens, 129
Mutations
   aflatoxin and, 77–78
   genotoxins, see Carcinogens
myc, 211
N-Methyl-D-aspartate (NMDA), 25

N

NAD(P)H, 48, 49, 52, 60
neu, 130, 131
Neuropathy target esterase (NTE)
      hypotheses, 11–13
Neurotoxicity
   injury and repair, 19–30
      MPTP and neurodegenerative
         disorders, 23–26
      organophosphates, 26–30
      pyrethroids, 21–23
   selectivity and mechanisms, 5–13
      delivery of intoxicant, 5
      organophosphorus-induced delayed
         neuropathy, 11–13
      pyrethroids, 8–11
      stages of, 4
      triorganotins, 5–8
Nigrostriatal tract, 24
NISH, see *In situ* hybridization
Nitrofurantoin, 49
NMDA antagonists, 25
N-ras, 77
Nuclear membrane, 48

O

Oligonucleotide probe preparation for *in situ*
      hybridization, 153
Oncogenes and proto-oncogenes, 130
   aflatoxin susceptibility, 77–78
   in apoptosis, 189, 211–213
Organophosphorus-induced delayed
      polyneuropathy (OPIDP), 11–13,
      26–30
Oxidative burst, 40
Oxygen toxicity, 35–60
   cell injury by redox cycling generation of
      oxygen species, 40–42
   extracellular calcium, 53–55
   fiber pathogenicity, 93–94
   glutathione, 47–49
      as cellular protective agent, 49–51
      mitochondrial, 55–56
      redox cycle, 51–53
   lipid peroxidation and antioxidant
      interactions, 42–49
      alpha-tocopherol-ascorbate
         relationship, 42–46
      glutathione, 47–49
      thiol groups, 47–50

mitochondrial glutathione, 55–56
mitochondrial permeability transition, 57–60
origin of oxidative stress, 37–40
pyridine nucleotide oxidation during permeability transition, 60

P

Paraquat, 40, 54
Parkinsonism, 24, 25, 26
Pentachlorophenol, 7
Permeability transition, 57–60
Permethrin, 9, 22
Peroxidation of lipids, 36
Pesticides
    pyrethroid neurotoxicity, 8–11, 21–23
    reactive oxygen species sources, 38, 40, 41
p53, 77, 78
Phagocytes, 40, 90
Phenolic hexachlorophane, 6
Phenols, acidic, 7
Phenylimino compounds, 6, 7, 8
Phenylmethanesulfonyl fluoride (PMSF), 12, 28
Phenyl-N-methyl-N-benzylcarbamate (PMBC), 28
Phosphate, and mitochondrial permeability transition, 57–59
Phosphoramido compounds, 12
Phosphorus-32 postlabeling, 105–108, 135
Polycyclic aromatic hydrocarbons, 102
Polymorphonuclear leukocytes, 40
Potassium channels, 21
Prednisone, 137
Procarbazine, 137, 138
Programmed cell death, see Apoptosis
Protein-carcinogen adducts, 111–116
    aflatoxin, 77
    albumin, 112–113
    hemoglobin, 113–116
Protein sulfhydryl groups, 39
Proto-oncogenes, see Oncogenes and proto-oncogenes
Pyrethroid neurotoxicity
    injury and repair, 8–11
    selectivity and mechanisms, 8–11

Q

Quinones, 42

R

Radioimmunoassays, 108
ras, 77, 102, 130
Redox cycling, cellular damage during, 36, 40–42
Redox cycling agents, 37
Research, see Animal models
Rhodesian chrysotile, 87
RNA, see *In situ* hybridization
Rockwool, 90, 92, 93
Ruthenium red, 54

S

Saligenin phosphate, 5
Semiquinone radical, 45
Sodium channels, 21, 22
Somatic mutation theory, 129
Species differences, aflatoxin susceptibility, 70–74
Strain differences, see Animal models, genetic variation in
Styrene, 102
Sulfhydryl groups, 39
Sulfhydryl reagents, and mitochondrial permeability transition, 57, 58
Superoxide, 37, 38
    in diquat toxicity, 40
    disproportionation of, 38
Superoxide dismutase, 38
Superoxide scavengers, 42

T

Tamoxifen, 133–134, 211
Thiol groups, 47, 48–49, 50
Thiols, see Glutathione
Thromboxane A2 agonists, 189
Thymocyte apoptosis, 189–199
Tin compounds, 5–8, 189
α-Tocopherol, 37, 42–47, 54
Toremifene, 134
Transition metals
    chelation, 42
    diquat toxicity, 40
    hydrogen peroxide-ferrous iron interaction, 42
    iron-lipid hydroxide-dependent pathway, 43–44
    and oxidative stress, 37
Tremolite, 88

Tremor syndrome, 9, 10, 21
Trifluroperadine, 56
Tri-*o*-cresyl phosphate, 5
Triorganotins
  and apoptosis, 189
  mechanisms of neuropathy, 5–8
T (tremor)-syndrome, 9, 10, 21
Tumor suppressor genes, 77–78

U

Ubiquinol, 56
Ubiquinone, 38
Ultrasensitive enzymatic radioimmunoassay
  (USERIA), 108
Unsaturated fatty acids, lipid peroxidation, 42
Urinalysis, 110–111

V

v-Ha-ras, 130
Vincristine, 137
Viral DNA preservation in archived tissues,
  149
Vitamin C, 37, 42–47
Vitamin E, 37, 42–47, 54
Voltage-dependent sodium channels, 21–23

X

Xanthine dehydrogenase, 55

Z

Zinc, and apoptosis, 190–195, 198, 199
Zirconia fiber, 91